Introduction to Spectroscopic Methods

THE AUTHOR

Dr. Timir Tripathi is the Regional Director of Indira Gandhi National Open University (IGNOU), Regional Centre Kohima, Nagaland, India. Earlier, he served as a Senior Assistant Professor at the Department of Biochemistry, North-Eastern Hill University, Shillong, India. He holds a Ph.D. from the Central Drug Research Institute, Lucknow, India. He was a visiting faculty at ICGEB, New Delhi, India, and Khon Kaen University, Thailand. He is known for his research in protein biophysics, biochemistry, structural biology, and drug discovery. He has over 17 years of experience in teaching and research on protein structure, function, and dynamics at the post-graduate and doctoral levels. He has developed and improved methods to investigate and analyze proteins. He has handled 12 research projects as a principal investigator from various national and international funding agencies. He has received several national and international awards. He is an elected member of the National Academy of Sciences, India, and the Royal Society of Biology, UK. He has published more than 100 research papers, reviews, commentaries, viewpoints, and editorial articles in international journals, has edited four books, and published several book chapters.

Introduction to Spectroscopic Methods

– Author –

Timir Tripathi

Regional Director, Indira Gandhi National Open University, Regional Centre Kohima, Kohima, Nagaland, India

Senior Assistant Professor, Department of Biochemistry, North-Eastern Hill University, Shillong, Meghalaya, India

2023

Daya Publishing House®
A Division of

Astral International Pvt. Ltd.
New Delhi – 110 002

ISBN: 9789359191492

Publisher's Note:

Every possible effort has been made to ensure that the information contained in this book is accurate at the time of going to press, and the publisher and author cannot accept responsibility for any errors or omissions, however caused. No responsibility for loss or damage occasioned to any person acting, or refraining from action, as a result of the material in this publication can be accepted by the editor, the publisher or the author. The Publisher is not associated with any product or vendor mentioned in the book. The contents of this work are intended to further general scientific research, understanding and discussion only. Readers should consult with a specialist where appropriate.

Every effort has been made to trace the owners of copyright material used in this book, if any. The author and the publisher will be grateful for any omission brought to their notice for acknowledgement in the future editions of the book.

Published by : **Daya Publishing House®**
A Division of
Astral International Pvt. Ltd.
– ISO 9001:2015 Certified Company –
4736/23, Ansari Road, Darya Ganj,
New Delhi-110 002
Ph. 011-43549197, 23278134
E-mail: info@astralint.com
Website: www.astralint.com

Foreword

Spectroscopic methods are the foundation techniques used to analyze the structure, function, and electronic properties of biomolecules. During the past few decades, biomolecular spectroscopy has made significant methodological advancements. The advances in technology, computational methods, and hardware have allowed the precise determination of structural and functional properties of biomolecules more rapidly, which was not possible to achieve earlier.

Spectroscopy is an indispensable tool to support all forms of laboratory research. This book provides concise notes on various spectroscopic methods. Explaining the fundamentals of spectroscopic methods is a tricky job. Particularly to the students of life science and related subjects, understanding the principles and applications of spectroscopy is tedious as it requires a knowledge of physics and chemistry. The primary purpose of this book is to foster a deeper understanding of the basic concepts and applications of spectroscopic methods. Each chapter lucidly outlines the fundamental principles, methods, and applications. The students will obtain an idea of the fundamentals and principles of the widely used spectroscopic methods for their applications in the field of biochemistry, biotechnology, molecular biology, pharmacology, pharmaceutics, biophysics, structural biology, and analytical and medicinal chemistry.

As a teacher, I have mentored a large number of young and dynamic students for the last 25 years. I consider this book a fantastic, comprehensive, and handy guide to contemporary spectroscopic methods. This book will benefit students and researchers as they progress in their fields. I congratulate Dr. Tripathi for bringing out this exceptional compendium for the benefit of students. I am confident that the book will be a promising and valuable addition to the bookshelf of students and the libraries of universities and colleges.

Prof. Prabha Shankar Shukla

Vice-Chancellor

North-Eastern Hill University, Shillong

Preface

Spectroscopy is the branch of science that studies the interaction between light and matter. Spectroscopic methods are widely used for quantitative and qualitative analysis of biomolecules in various branches of modern biology, life sciences, chemistry, physics, and pharmacology. In spectroscopy, the spectrum produced by matter when it interacts with or emits light is measured. A clear understanding of the underlying principles of spectroscopy is vital to learning their applications.

This book presents a series of lecture notes on various spectroscopic methods used in biological and chemical sciences. The students will obtain an idea of the fundamentals and principles of the widely used spectroscopic methods for their applications. The chapters have been meticulously crafted, so students can understand the concept easily and get the best out of each chapter. The book will serve as a guiding light for students and researchers equally, who are engaged in the field of biochemistry, biotechnology, life sciences, chemistry, medicinal chemistry, analytical chemistry, biophysics, physics, pharmacology, *etc.* It will help the readers understand the concepts, principles, and applications of biomolecular spectroscopy. Any student studying spectroscopy in his course syllabus would require this book, regardless of whether the student is an undergraduate or postgraduate or even an independent researcher. This book will be extremely handy for all.

The 12 chapters cover individual methods that are useful in biological and chemical sciences. The basic principles of spectroscopy are discussed in an introductory chapter. Techniques such as UV/Vis spectroscopy, fluorescence spectroscopy, circular dichroism spectroscopy, infra-red spectroscopy, Raman spectroscopy, NMR spectroscopy, X-ray spectroscopy, electron spin resonance spectroscopy, atomic absorption spectroscopy, nephelometry and turbidimetry are discussed. In the end, there is a chapter on mass spectrometry.

The chapters are written in simple language with clear illustrations and figures that the students can easily understand. They provide an in-depth understanding of the theory behind the techniques, with each chapter discussing the basics, principles, associated concepts, and applications of the method.

I want to express my sincere gratitude to many of my friends and colleagues who helped review the book and provided helpful criticism and comments: Krishnananda Chattopadhyay (CSIR-IICB, Kolkata), Krishna Mohan Poluri (IIT Roorkee), Sudhir Kumar Singh (IMS BHU, Varanasi), Sanjit Kumar Roy (VIT, Vellore), Dev Bukhsh Singh (Siddharth University, Kapilvastu), Aditya Kumar Padhi (IIT BHU, Varanasi), and Prakash Saudagar (NIT Warangal). I also acknowledge my PG and PhD students at the North-Eastern Hill University, Shillong, to whom I taught biochemical techniques for over 13 years, which helped me to develop this book. Finally, I would like to thank my wife, Tripti, for her constant support and patience in providing me with time so that I could complete this book.

Timir Tripathi

Kohima, Nagaland

Contents

Foreword *v*

Preface *vii*

1. Fundamental Principles of Spectroscopy **1**

1. *Introduction*
2. *Basic Features of Light*
3. *Electromagnetic Spectrum*
4. *Interaction of Light with Matter*
5. *Atomic/Molecular Processes Associated with Light*
6. *Basics of Electronic Transitions*
7. *General Construction of a Spectrometer*
8. *Physical Limitations of a Spectroscopic Measurement*

2. Ultraviolet/Visible Spectroscopy **15**

1. *Introduction*
2. *Chromophores*
3. *Transmittance and Absorbance*
4 *Principle of UV/Vis Spectroscopy-Beer-Lambert's Law*
5. *Construction of a Spectrophotometer*
6. *Types of Spectrophotometers*
7. *Application of UV/Vis Spectroscopy*

3. Fluorescence Spectroscopy **33**

1. *Introduction*
2. *Principle of Fluorescence Spectroscopy*
3. *Important Terminologies in Fluorescence Spectroscopy*
4. *Effect of Solvents on Fluorescence Spectra*
5. *Characteristics of the Fluorescence Phenomenon*
6. *Fluorophores*
7. *Construction of a Fluorescence Spectrometer*
8. *Applications of Fluorescence Spectroscopy*

4. Circular Dichroism Spectroscopy **55**

1. *Introduction*
2. *Plane Polarized Light*
3. *Circularly Polarized Light*
4. *Principle of Circular Dichroism Spectroscopy*
5. *Construction of a CD Spectrometer*
6. *CD of Biomolecules*
7. *Applications of CD spectroscopy*

5. Infra-red Spectroscopy **69**

1. *Introduction*
2. *Molecular Vibrations*
3. *Absorption of IR Radiation*
4. *Characteristics of IR Spectra*
5. *Construction of an IR Spectrometer*
6. *Applications of IR Spectroscopy*

6. Raman Spectroscopy **79**

1. *Introduction*
2. *Scattering of Light*
3. *Origin of Raman Spectra*
4. *Principle of Raman Scattering*
5. *Construction of a Raman Spectrometer*
6. *Improving Raman Signals*
7. *Applications of Raman Spectroscopy*
8. *Differences between IR and Raman Spectroscopy*

7. Nuclear Magnetic Resonance Spectroscopy **95**

1. *Introduction*
2. *Principle of NMR Spectroscopy*
3. *Working of an NMR Spectroscopy Experiment*
4. *^{13}C NMR Spectroscopy*
5. *^{1}H NMR Spectroscopy*
6. *Example of a ^{13}C and ^{1}H Spectra*
7. *Applications of NMR*

8. X-ray Spectroscopy **111**

1. *Introduction*
2. *Properties of X-rays*
3. *Steps in X-ray Crystallography*
4. *Construction of an X-ray Spectrometer*
5. *Working of an X-ray Diffractometry Experiment*
6. *Applications of X-ray Spectroscopy*
7. *Comparison of X-ray Crystallography and NMR Spectroscopy*

9. Electron Spin Resonance Spectroscopy **121**

1. *Introduction*
2. *Principle of ESR Spectroscopy*
3. *Construction of an ESR spectrometer*
4. *Applications of ESR Spectroscopy*

10. Atomic Absorption Spectroscopy **125**

1. *Introduction*
2. *Principle of Atomic Absorption Spectroscopy*
3. *Steps in AAS*
4. *Instrumentation*
5. *Experimental Procedures*
6. *Applications of AAS*

11. Turbidimetry and Nephelometry **131**

1. *Introduction*
2. *Turbidimetry*
3. *Nephelometry*

4. *Factors Affecting Light Scattering*
5. *Applications of Turbidimetry and Nephelometry*

12. Mass Spectrometry **135**

1. *Introduction*
2. *Principle of Mass Spectrometry*
3 *Construction of a Mass Spectrometer*
4. *Types of Mass Spectrometers*
5. *Mass Spectrometry Coupled with Chromatography*
6. *Protein Sequence Identification by Mass Spectrometry*
7. *Applications of Mass Spectrometry*
8. *Difference between NMR Spectroscopy and Mass Spectrometry*

Index **155**

Chapter 1

Fundamental Principles of Spectroscopy

1. Introduction

- ☆ To perform any measurement, we must observe the interaction of the sample of interest with another particle. When light is used as the interacting particle, the measurement is called spectroscopy.
- ☆ When light passes through a sample, it interacts with the sample. Hence, if we monitor the changes in the light coming out from the sample, we may learn something about the sample. This process is the basis of spectroscopy.
- ☆ Spectroscopy is the measurement of light absorbed (absorption spectroscopy) or emitted (emission spectroscopy) when a molecule/ atom/ion present in a sample moves from one energy state to another energy state.
- ☆ Spectroscopic methods use light to interact with matter and thus obtain information on certain features of the sample.

Note

- ☆ **Spectro**: light; **-scope**: looking, examining; **-graph**: recording; **-meter, -metry**: measuring.
- ☆ **Spectroscopy**: Study of the interaction of light with matter.
- ☆ **Spectrometry:** Quantitative measurement based on information from a spectrum.

- **Spectrum (*plural* Spectra):** Display of the intensity of radiation emitted, absorbed, or scattered by a sample *vs* a quantity related to photon energy (such as wavelength or frequency).
- **Spectrometer:** An instrument that measures the differences in various properties caused by a sample over a particular range.
- **Spectrophotometer:** It is a spectrometer that measures the absorption or emission of light from a sample.

2. Basic Features of Light

- Light or electromagnetic radiation is a form of energy that can be described in two ways:
    1. **As a particle:** Light behaves as a stream of energy pulses, photons, travelling as packets of energy at the velocity of 3×10^8 m/sec (*i.e.*, the speed of light in a vacuum). Thus, light has a particle-like property.
    2. **As a wave:** Light behaves as electromagnetic waves, as electric and magnetic fields oscillating in space and time at the speed of light. Thus, light has a wave-like property.
- From the above two definitions, it is apparent that light has a dual nature (wave-particle duality), *i.e.*, light has properties of a wave and a particle (with zero mass).

Note

- The optical properties of light, such as diffraction or refraction, can be best explained by thinking of light as a wave.
- However, the interaction of light and matter (*i.e.*, absorption and emission processes) can be better described if the light is treated as a particle or photon.
- Nonetheless, both definitions provide a useful description of light.
- The velocity of light is 3×10^8 m/sec in a vacuum; in a medium other than the vacuum, the velocity decrease.

- Electromagnetic radiation (EMR), *i.e.*, light, can be described as an electromagnetic wave propagating in space in a linear path and at a constant velocity (**Figure 1**).
- The electric and magnetic field vectors are directed perpendicular to each other.
- Light photons have an energy that is defined by the wavelength of light.

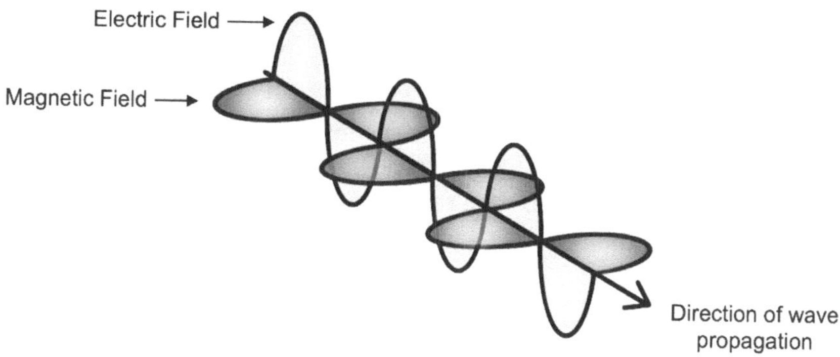

Figure 1: An Electromagnetic Wave Propagating in a Direction.
The direction is perpendicular to the plane where the electric field and magnetic field lie.

> ***Note***
> - For UV/Visible, circular dichroism (CD), fluorescence, and infra-red (IR) spectroscopy, the focus is on the electric field vector.
> - Electron paramagnetic resonance (EPR) and nuclear magnetic resonance (NMR) spectroscopy focus on the magnetic field vector.

- The wavelength (λ) is the distance between two consecutive peaks (one complete cycle) of the wave and is measured in meters or nanometers (nm) (**Figure 2**).

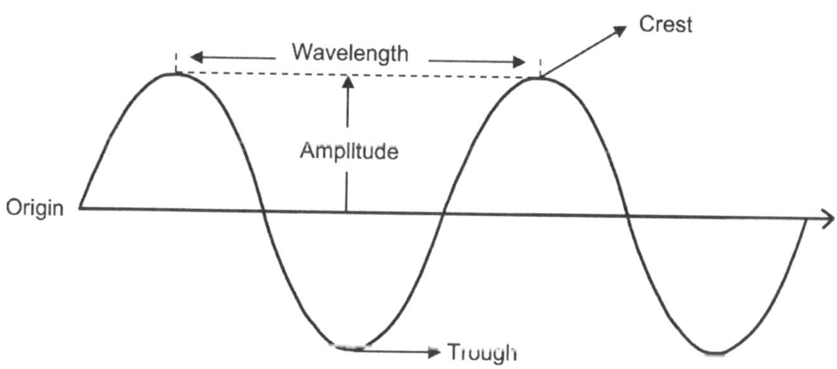

Figure 2: A Wave Traversing through Space.
The highest point of a wave is called the crest, and the lowest point is the trough. The vertical distance of the crest/trough is the amplitude. The horizontal distance between two adjacent crests or troughs is known as the wavelength.

- The maximum displacement of the wave is called the amplitude, *i.e.*, the distance between the origin and crest (or trough).

☆ Two other parameters should also be noted:

1. Wavenumber (ν): describes the number of wave cycles per unit distance and is typically measured in cm^{-1} or m^{-1}.
2. Frequency (ν) of the EMR is the number of waves that pass a point per unit time (*i.e.*, within the timeframe of 1 s). It, therefore, has units of $1\ s^{-1} = 1\ Hz$. The relationship between the frequency and the wavelength is defined by $c = \lambda\nu$.

3. Electromagnetic Spectrum

☆ The electromagnetic spectrum is the classification of electromagnetic rays according to wavelength or frequency. The wavelength, energies, and energy transition associated with various electromagnetic waves are shown in **Figure 3**.

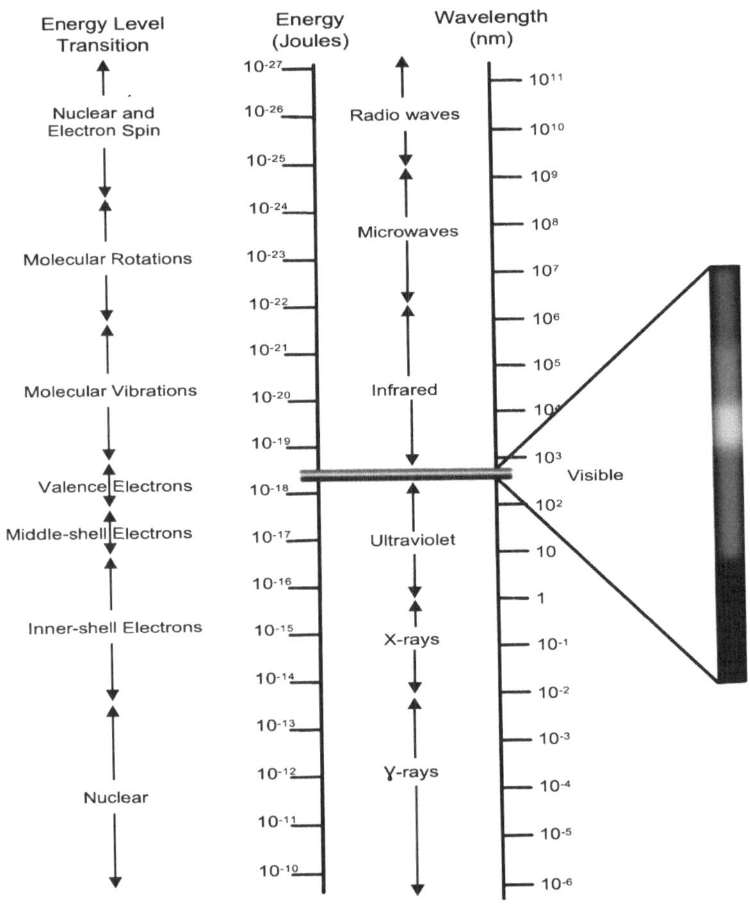

Figure 3: Electromagnetic Spectrum.

The wavelength, energies, and energy transition associated with various electromagnetic waves are shown.

- Although all electromagnetic rays travel at the speed of light, they have different wavelengths, frequencies, and energies.
- The electromagnetic spectrum comprises all electromagnetic radiations and consists of many subranges, such as UV, visible, X-rays, gamma rays, *etc.*
- There are no precise boundaries between any of these contiguous subranges, so the ranges tend to overlap.
- The electromagnetic spectrum includes all radio waves (*e.g.*, commercial radio and television, microwaves, radar), UV, visible, IR, X-rays, and gamma rays.
- Almost all wavelengths of light can be used for spectroscopy.

4. Interaction of Light with Matter

- An atom can be described as certain discrete states called quantum states. Each quantum state is associated with definite energy.
- Atoms/molecules can absorb or emit energy in discrete packets called quanta (singular: quantum).
- Atoms/molecules can possess energies in different forms, such as electronic, vibrational, rotational, translations, and spin orientation energies (vibrational energies are quite small). The total energy is the sum of all these components.
- Absorption associated with electronic transitions occurs in a femtosecond (10^{-15} seconds) timescale.
- The atomic energy levels are measured by observing transitions between two levels. For *e.g.*, an atom in its lowest energy state (ground state) can be excited to a higher state only if the energy provided is equal to the energy difference between the two levels. Thus, by measuring the energy of the light that the atom has absorbed, the difference in its energy levels can be determined (**Figure 4**).

Note

- The interaction of light with matter can be explained by quantum mechanics.
- Quantum mechanics can be understood by comparing it with Newtonian mechanics.
- Newtonian mechanics relates to the motions of large particles, while quantum mechanics relates to the motions of small particles (*e.g.*, atoms and molecules).
- Newtonian mechanics considers the particle's motion as a continuous entity, while quantum mechanics considers that small particles exist in discrete states of motion (energies).

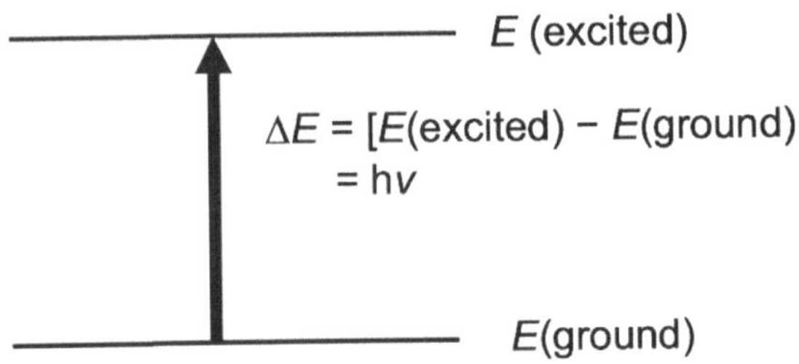

Figure 4: The Excitation Process.

The orbiting electrons in a ground state can absorb the excitation energy to jump to a higher excited state energy level.

- ✰ When a particle interacts with light, it may absorb light and undergo a change in energy.
- ✰ In order to interact with light, the molecules must have electric or magnetic properties that could be influenced by the electric or magnetic components of the light.
- ✰ To understand the interaction of light and matter, imagine light as a particle or photon instead of a wave.
- ✰ When a photon interacts with a sample, its energy is transferred into the sample.
- ✰ The atoms/molecules of the sample exist in many defined energy states. Once the photon's energy is transferred into the sample, there is a change in the electronic state of the sample, *i.e.*, its electrons may jump to higher orbits.
- ✰ The energy of a photon absorbed or emitted during a transition from one energy level to another is given by:

$$E = h\nu$$

where h is the Planck constant ($h = 6.63 \times 10^{-34}$ Js) and ν is the frequency of the photon. Now since

$$c = \nu \lambda$$

therefore,

$$E = \frac{hc}{\lambda}$$

- ✰ Thus, radiation with a shorter wavelength has higher energy and vice versa. Hence, the amount of energy a light photon carries depends on wavelength, *i.e.*, the higher the wavelength, the lower the energy of the photon and vice versa.
- ✰ To promote an electron to a higher level (excited state), energy must be added to the system. The electron jumps from the ground state (S_0) to the excited state (S_1), giving rise to an absorption spectrum.

- ☆ Subsequent relaxation of the molecule from S_1 to S_0 will occur, which gives rise to the emission spectrum.

Note

- ☆ For non-fluorescent molecules, the relaxation is accompanied by the emission of heat (ΔH).
- ☆ For fluorescent molecules, the relaxation is accompanied by the emission of light in the form of fluorescence (discussed later in fluorescence spectroscopy).
- ☆ The photon and wave theories are linked by Plank's law ($E= hc/\lambda$).

- ☆ All electromagnetic radiations have the same velocity (that of light- 3×10^8 m/sec) but have different wavelengths. This means that X-rays, UV rays, visible light, radio waves, and all the other parts of the electromagnetic spectrum are fundamentally the same thing - electromagnetic radiation/light; the difference is in the energy (or wavelength) of these electromagnetic radiations.
- ☆ The energy of electromagnetic radiation depends inversely on the wavelength of the radiation (**Figure 5**). For *e.g.*, an X-ray photon has a low wavelength and thus high energy, while a radio wave photon has a high wavelength and, therefore, low energy.

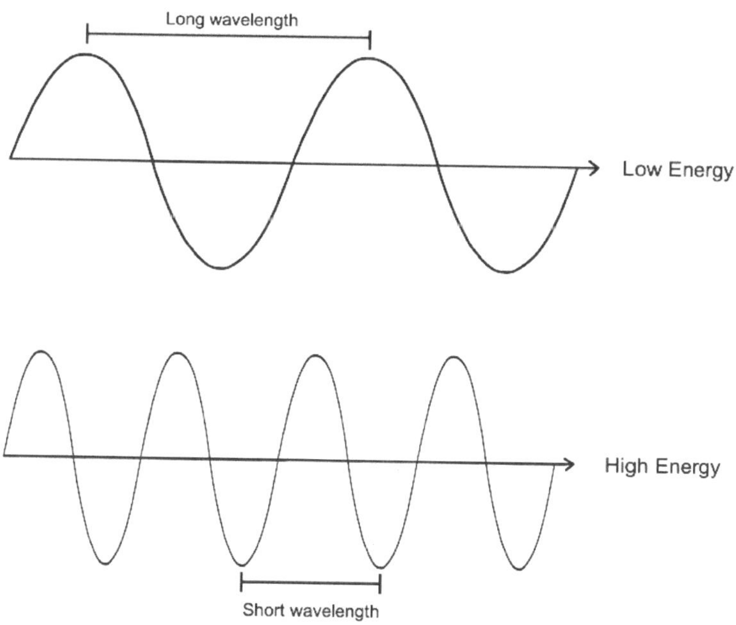

Figure 5: A Representation of Waves Showing the Relation between Energy and Wavelength. The longer the wavelength, the lower the energy and vice versa.

- ☆ The energy of the photon determines the type of transition or interaction that occurs.
- ☆ The photons of the visible region contain energies of ~36-72 kcal/mol, while near the UV region, the energy range is ~140 kcal/mol. Thus, photon energies for the UV region are higher than that of the Vis region.
- ☆ UV rays of wavelength less than 200 nm are challenging to handle and rarely used as a tool for structural analysis (if at all used, only under vacuum/inert gas).
- ☆ All molecules absorb light of only a certain wavelength and transmit the rest.
- ☆ When light falls upon a sample (interacts with matter), the light can be reflected, refracted, transmitted, absorbed, or diffracted **(Figure 6)**. In scattering, the radiation is first absorbed and then instantaneously re-emitted entirely and uniformly in all directions but otherwise is unchanged. The light can also excite fluorescence or cause a photochemical reaction.

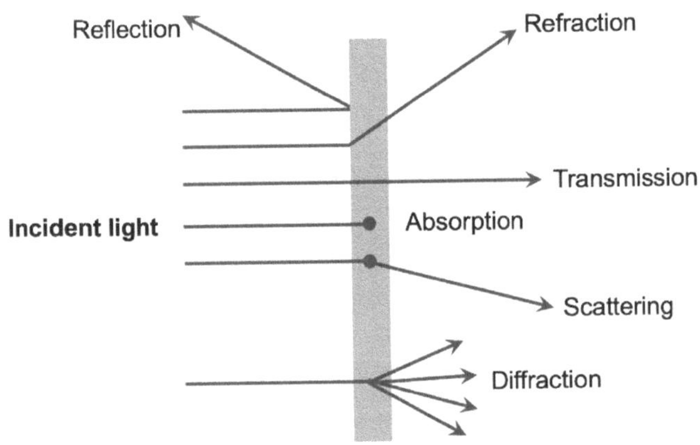

Figure 6: Interaction of Light with a Sample.

When an incident light encounters a partially transparent sample, it is either reflected, refracted, transmitted, absorbed, scattered, or diffracted.

- ☆ In fluorescence, a photon is first absorbed and the molecule is excited to a higher energy state. The molecule then returns to an intermediate energy level by re-emitting a photon. Since the molecule retains some energy of the incident photon, the emitted photon has less energy and hence a longer wavelength than the absorbed photon (discussed in chapter 3).
- ☆ Absorption spectroscopy is related to the processes of absorption and transmission.

- The conditions under which the sample is studied are carefully maintained to keep reflection, scattering, and fluorescence to a minimum.
- UV/Vis spectroscopy is almost entirely used for quantitative analysis, *i.e.*, estimating the amount of a compound present in the sample, as the data is insufficient to identify an unknown sample.
- In general, while measuring the UV/Vis spectrum, one wants that only absorbance should occur.
- Since light is a form of energy, absorption of light by matter causes the energy content of the atoms/molecules to increase.
- The total potential energy of a molecule is the sum of its electronic, vibrational, and rotational energies:

$$E_{total} = E_{electronic} + E_{vibrational} + E_{rotational}$$

Note

- Coloured substances have absorption spectra in the visible region.
- Colorimetry is a method used to determine the concentration of coloured compounds in solutions using the Beer-Lambert law.
- Many substances that appear colourless may also have absorption spectra. In this case, the absorption occurs in the IR or UV, not in the visible region.

5. Atomic/Molecular Processes Associated with Light

- When atoms/molecules absorb light, they undergo rotation, vibration, electronic transitions, and nuclear transitions. The energies associated with these processes correspond to different wavelengths of light. The energy of the photon determines the type of transition or interaction that occurs.
- The energy states of ground/excited orbitals are separated by a large gap (~eV) compared to the vibration and rotation energy levels.
- A single ground/excited state orbital may have several vibration and rotation energy levels.
- Thus, electronic transitions occur between different electronic states, while vibrational transitions occur in the same electronic state, and rotational transitions occur in the same vibrational state.
- The electronic and vibrational energy levels can be understood by analogy to a "Stadium".
- Each electronic energy level (orbital) is composed of several vibrational energy levels.
- The steps between the vibrational energy levels are much smaller than the steps we see between electronic energy levels (*i.e.*, orbitals). Thus,

less energy is required for transitions between vibrational energy levels than for electronic transitions. Imagine the decks as the energy levels of orbitals and the rows as the vibrational energy levels.

- ☆ The UV spectra appear broad because the value of ΔE for the transition also depends on the difference between vibrational energy levels.
- ☆ Various electronic orbitals or energy levels are shown in **Figure 7**. The electronic transitions involve two energy potential wells.

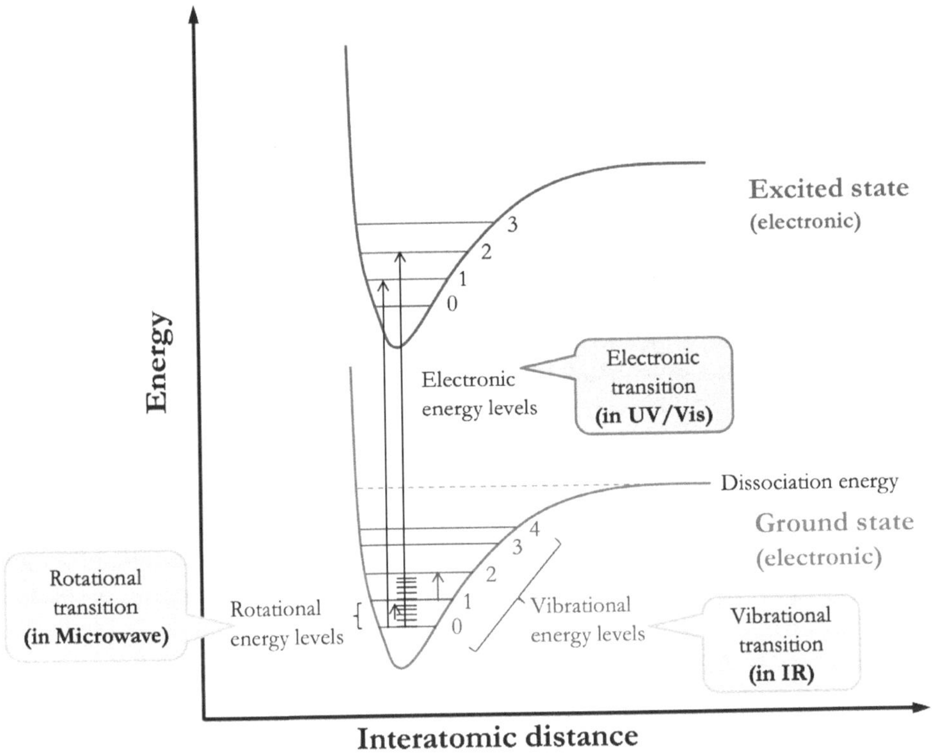

Figure 7: Various Electronic Orbitals (energy levels).
The electronic transitions involve two energy potential wells.

5.1. Radio Waves

- ☆ The radiofrequency region has very low energies that correspond to the energy differences in the nuclear and electron spin states. These frequencies, therefore, find applications in nuclear magnetic resonance and electron paramagnetic resonance spectroscopy.

5.2. Microwaves

- ☆ Microwaves have energies between radio and infra-red waves and find applications in rotational and electron paramagnetic resonance spectroscopy.

5.3. Infra-red Rays

- The energies associated with the infra-red region are not large enough to excite electrons. When molecules absorb energy in the infra-red region, they undergo molecular vibrations, and vibrational excitation of covalently bonded atoms or groups occurs. Thus, infra-red spectroscopy is also known as vibrational spectroscopy. It is a very useful technique for functional group identification in organic compounds.

5.4. UV/Visible Rays

- UV and visible rays induce electronic transitions in the molecules. Therefore, the spectroscopic methods using UV or visible light come under 'electronic spectroscopy'. Such as UV/Vis, circular dichroism, and fluorescence spectroscopy.

5.5. X-rays

- X-rays are high-energy electromagnetic rays and cause transitions in the internal electrons of the molecules.

6. Basics of Electronic Transitions

- The colour of any substance is related to its electronic structure.
- Molecules containing π-electrons or non-bonding electrons (n-electrons) can absorb energy in the form of UV/Vis light to excite these electrons to higher anti-bonding molecular orbitals.
- Thus, the light absorption in the UV or Vis region is accompanied by a change in the electronic state of the molecules in the sample. The light energy promotes electrons from their ground state orbitals to higher energy, excited state orbitals, or anti-bonding orbitals. These energies in the UV/Vis region are sufficient to promote or excite electrons to a higher energy orbital.
- Generally, three types of ground state orbitals may be involved in electronic transitions: σ (bonding), π (bonding), and n (non-bonding) atomic orbital. Besides, two types of anti-bonding orbitals may be involved in the transition: σ * (sigma star) orbital and π* (pi star) orbital **(Figure 8)**.

Note

- There is nothing as an n* anti-bonding orbital as the n electrons do not form bonds).

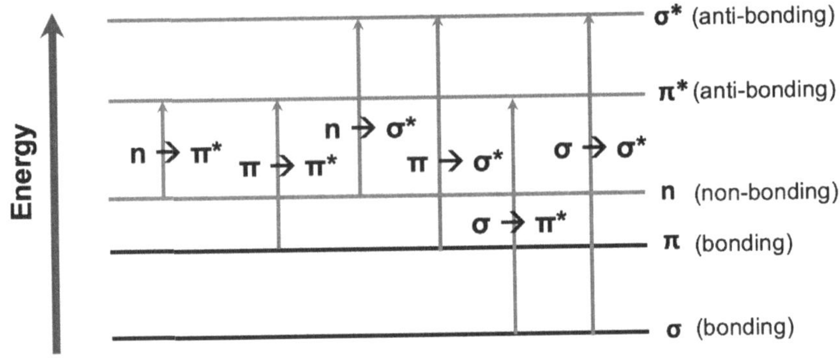

Figure 8: Energy Scheme for Molecular Orbitals.

The arrows represent possible electronic transitions. The length of the arrows represents the amount of energy required that is required for the particular transition.

- ☆ Of the six transitions shown, only the two lowest energy ones (left two) are achieved by the energies available in the UV light.
- ☆ As a rule, molecules absorb UV/Vis energy to promote their electrons to higher vacant anti-bonding orbitals having low vacant molecular orbitals, resulting in an excited state.

6.1. $\sigma \rightarrow \sigma^*$ Transitions

- ☆ An electron in a bonding s-orbital is excited to the corresponding anti-bonding orbital.
- ☆ A very high amount of energy is required for such transitions.
- ☆ *E.g.*, methane (having four C-H bonds) can only undergo $\sigma \rightarrow \sigma^*$ transitions) and shows an absorbance maximum at 125 nm.
- ☆ Absorption maxima due to $\sigma \rightarrow \sigma^*$ transitions are not observed in common UV/Vis spectra (200-700 nm)

6.2. n $\rightarrow \sigma^*$ Transitions

- ☆ Saturated compounds that contain atoms with lone pairs (non-bonding electrons) can undergo n $\rightarrow \sigma^*$ transitions.
- ☆ These transitions require less energy than $\sigma \rightarrow \sigma^*$ transitions and can be caused by light with a wavelength in the range of 150-200 nm.
- ☆ The number of functional groups with n $\rightarrow \sigma^*$ peaks in the UV region is very few.

6.3. n $\rightarrow \pi^*$ and $\pi \rightarrow \pi^*$ Transitions

- ☆ Most spectroscopy of organic compounds is based on transitions of n or π electrons to the π^* excited state.

☆ These transitions fall in an experimentally suitable region of the spectrum (190-380 nm for UV and 380-750 nm for the visible region).

☆ These transitions require an unsaturated group in the molecule to provide the π electrons.

☆ Generally, transitions obtained for $\pi \rightarrow \pi^*$ are stronger (higher energy) than $n \rightarrow \pi^*$ (lower energy).

7. General Construction of a Spectrometer

☆ The generation and analysis of spectra usually require the following instrument components in a spectrometer instrument, as shown in **Figure 9**:

1. Light source: energy source that can be input into the sample,
2. Wavelength selector: separates the light into its component wavelengths,
3. Detector: detects the light signal,
4. Computer: processes and displays the signal as an output spectrum.

☆ The instrument used to produce the light, separate it into its component wavelengths, and detect the spectrum after absorption/emission, is called a spectrometer.

☆ The spectra obtained are either absorption spectra (having a continuously bright background) or emission spectra (having one or more bright lines or bands on a dark background).

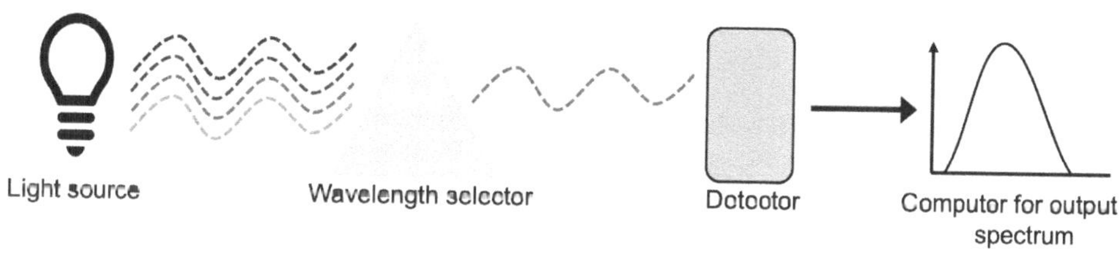

Figure 9: Main Components of a Spectrometer Instrument.
Details are discussed in the main text.

8. Physical Limitations of a Spectroscopic Measurement

☆ Spectroscopic methods offer the necessary tools to investigate the structure and function of biomolecules.

☆ A given spectroscopic method might be suitable for investigating some compounds (or analytes) and answering particular questions, but it may not be useful for others.

☆ To assess the usefulness of a given spectroscopic method, it is necessary to identify its limitations with respect to its sensitivity, selectivity, and resolution.

1. Highly sensitive methods allow the study of small quantities/concentrations of a sample and yield enough information to distinguish multiple compounds. The most sensitive methods available today allow the observation of single molecules (by fluorescence or mass spectroscopy) but provide only limited information about a sample.
2. Selectivity allows the identification and characterization of a specific sample in the presence of others and can only provide the integrated signal of indistinguishable species. However, the scope of the method is limited if only a small set of samples is available for analysis.
3. Resolution is a primary issue in assessing the information offered by various spectroscopic methods and is limited by Heisenberg's uncertainty principle.

☆ To assess the usefulness of a given method, we have to consider the relationship between the question to be investigated, the spectroscopic method, and the limitations of the method.

Note

☆ **Sensitivity:** the ability to detect small quantities of a compound.

☆ **Selectivity:** the ability to distinguish or spatially resolve different compounds.

☆ The sensitivity and selectivity are characteristics of a particular method and determine whether this method is suited for the desired measurement.

☆ **Resolution:** the minimum difference that can be separated/identified by a method. The signals from different species should be distinguishable.

Chapter 2

Ultraviolet/ Visible Spectroscopy

1. Introduction

- ☆ The absorption of ultraviolet (UV) and visible rays is one of the most routinely used analytical methods.
- ☆ UV/Vis spectroscopy is a type of absorption spectroscopy in which light from the ultra-violet region (190-380 nm) or visible region (380-750 nm) is absorbed by a molecule resulting in the excitation of electrons from the ground state to a higher energy state.
- ☆ When a chemical compound absorbs UV/Vis light, it produces a distinct spectrum which helps in the identification of the compound.
- ☆ UV/Vis regions of the electromagnetic spectrum and their associated techniques are the most widely used analytical methods in biology.
- ☆ The electronic transitions in molecules can be classified according to the participating molecular orbitals.
- ☆ Out of the four possible transitions ($\sigma \rightarrow \sigma^*$, $n \rightarrow \sigma^*$, $n \rightarrow \pi^*$, and $\pi \rightarrow \pi^*$ transitions), only two can provide information upon interaction with light from the UV/Vis spectrum for biomolecules: $n \rightarrow \pi^*$ and $\pi \rightarrow \pi^*$.
- ☆ The energy associated with $\sigma \rightarrow \sigma^*$ and $n \rightarrow \sigma^*$ transitions are not within the range of UV/Vis spectroscopy and require higher energies.

Note

- ☆ UV region- 190-380 nm [$\pi \rightarrow \pi^*$ - stronger (higher energy, lower wavelength)].
- ☆ Visible region- 380-750 nm [$n \rightarrow \pi^*$ - weaker (lower energy, higher wavelength)].

- ☆ The aromatic systems with π electrons in conjugation absorb strongly in the UV range.
- ☆ Interestingly, the greater the length of a conjugated system in a molecule, the nearer the λ_{max} comes to the visible region.
- ☆ Thus, the particular energy required for a transition (and hence the wavelength of absorption) is a property of atoms/molecules rather than the electrons themselves.

2. Chromophores

- ☆ The part of the molecule responsible for light absorption is called a chromophore.
- ☆ A chromophore is a molecular group usually containing an unsaturated group (π bonds).
- ☆ The presence of conjugated double bonds in organic molecules results in an extended π-system of electrons which lowers the energy of the π^* orbital through electron delocalization.
- ☆ These systems possess $\pi \rightarrow \pi^*$ transitions in the UV/Vis range of the EM spectrum.
- ☆ Such molecules are beneficial in spectrometric applications.
- ☆ If a chromophore is conjugated with another (of the same type or a different type), then multiple chromophores are formed, having a new absorption band that is more intense and at a longer wavelength.

Note

- ☆ **Figure 1** shows the various terms associated with a spectrum.

  Bathochromic shift: Shift of an absorption maximum towards a higher wavelength.

  Hypsochromic shift: Shift of an absorption maximum towards a smaller wavelength.

  Hyperchromic shift: Increase in the intensity of the absorption maximum.

  Hypochromic shift: Decrease in the intensity of the absorption maximum.

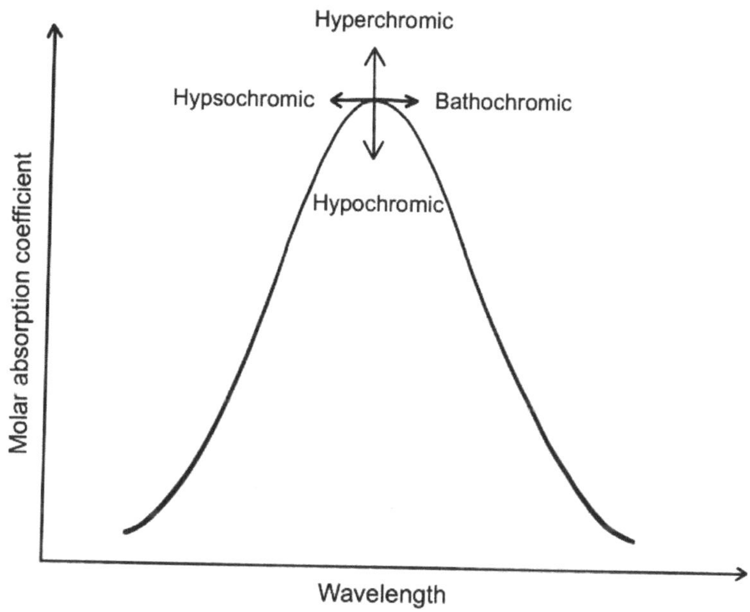

Figure 1: Terminology for Shifts in an Absorption Spectrum.

2.1. Chromophores in Proteins

- Proteins that contain coloured prosthetic groups (the non-protein components as chromophores) can be detected within the range of visible light (using the naked eye), such as haemoglobin, melanin, *etc.* However, most proteins are colourless and do not contain such groups. The colourless proteins can be detected and identified using UV spectroscopy.
- Protein contains three types of chromophores relevant to UV/Vis spectroscopy:
  - A) Aromatic amino acid side chains (tryptophan, tyrosine, phenylalanine),
  - B) Certain prosthetic groups and co-enzymes (*e.g.*, NADH, NADPH, FMN, FAD porphyrin groups as in haem, chlorophylls, *etc.*).
  - C) Peptide bonds (amide bonds),
- Various protein and non-protein chromophores are shown in **Table 1** and **Figure 2**.

Table 1: Various Protein and Non-protein Chromophores

Chromophore	*Molar Extinction Coefficient (ε_{max})*	*Absorption Maximum*	*Wavelength Range*
Tryptophan	~5700 $M^{-1}cm^{-1}$	280 nm	Near UV
Tyrosine	~1400 $M^{-1}cm^{-1}$	274 nm	
Phenylalanine	~200 $M^{-1}cm^{-1}$	250 nm	
Disulfide bond	~300 $M^{-1}cm^{-1}$	250-270 nm	
Peptide bond	~7000 $M^{-1}cm^{-1}$ (due to strong $\pi \rightarrow \pi^*$ transitions) ~190 $M^{-1}cm^{-1}$(due to weak $n \rightarrow \pi^*$ transitions)	190 nm 210-230 nm	Far UV
FAD	~11300 $M^{-1}cm^{-1}$	438 nm	Visible
NADH	~6300 $M^{-1}cm^{-1}$	340 nm	
NAD^+	~6300 $M^{-1}cm^{-1}$	260 nm	
Haemoglobin	~55500 $M^{-1}cm^{-1}$ ~54500 $M^{-1}cm^{-1}$	575 nm (Oxy-Hb) 555 nm (Deoxy-Hb)	

- Molecules such as FAD, NADH, and NAD^+ are important co-enzymes involved in electron transfer reactions (redox reactions). They are coloured. The protein containing these molecules can be readily identified using their absorption peaks in the visible region or visualized using their specific colour with the naked eye.

2.2. Chromophores in Nucleic Acids

- Nucleic acids show strong absorption in the 240-270 nm; the absorption peak is around 260 nm. However, depending on the base composition and environment, the peak can shift from 255 to 265 nm.
- The absorption of UV light by nucleic acids arises from $n \rightarrow \pi^*$ and $\pi \rightarrow \pi^*$ transitions of the purine (A, G) and pyrimidine (C, T, U) bases.
- The bases can be protonated. As a result, the absorption spectra are sensitive to the pH of the medium. The absorption spectra are also greatly influenced by electronic interactions between bases.

2.3. Auxochromes

- Auxochromes are functional groups that modify the absorption of neighbouring chromophores.
- They often have lone pairs (non-bonding electrons), *e.g.*, OH, -OR, $-NR_2$, and -X.
- They do not absorb but affect the spectrum of the chromophore to which it is attached.

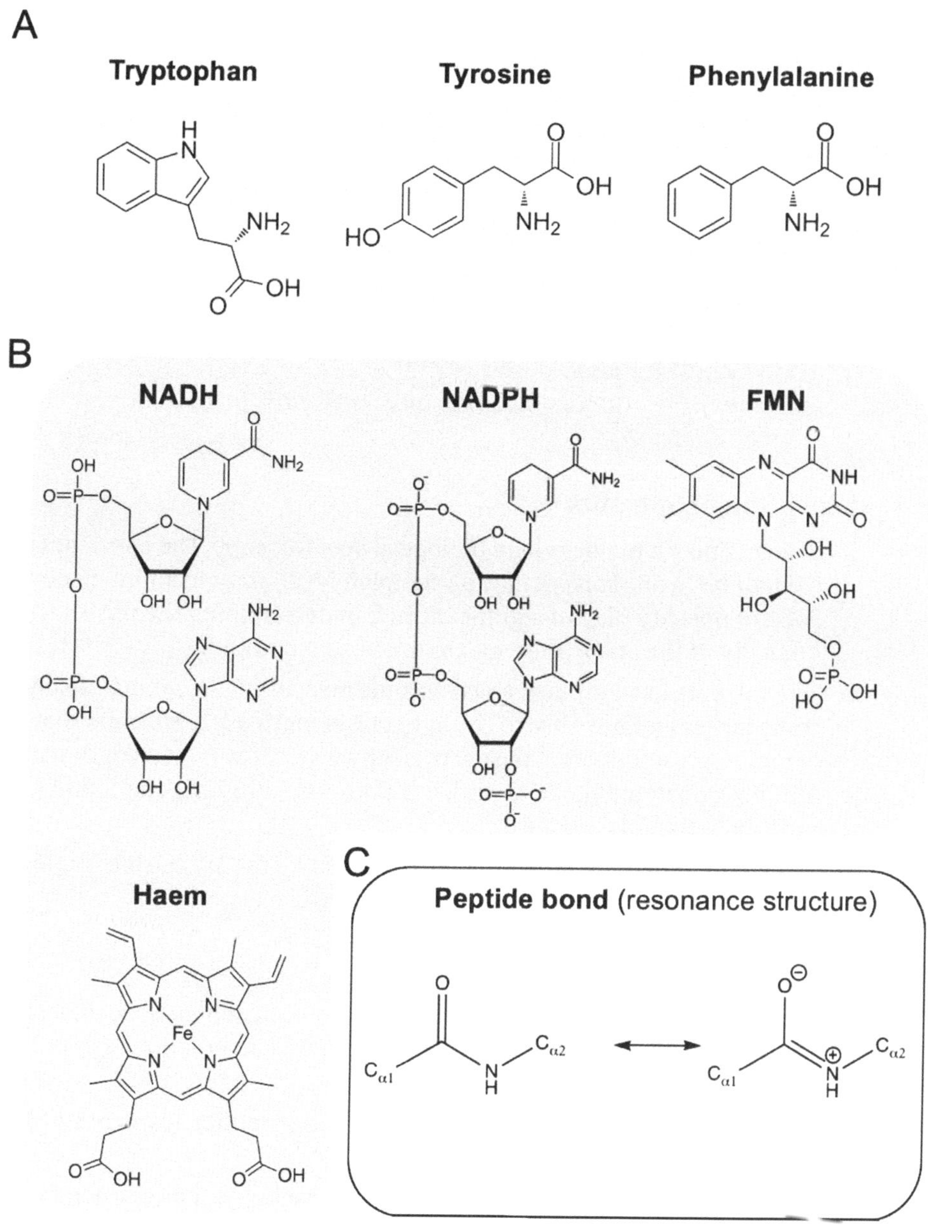

Figure 2: Molecular Structures of some Protein and Non-protein Chromophores. (A) Aromatic amino acids, (B) Prosthetic groups and co-enzymes, and (C) A peptide bond (the resonance structure of a peptide bond is shown).

☆ If an auxochrome is attached to a chromophore, it results in a bathochromic shift, as it extends conjugation and resonance.

Note

- Complex organic and inorganic molecules can interact with certain transition element metals and then absorb in the visible region, thereby helping in quantitative analysis. The absorptions are due to the movement of electrons between energy levels of the organo-metal complex.
- These complex systems are termed spectrometric reagents. The most common ones are dithizone, azo reagents, dithiocarbamate, 8-hydroxyquinoline, formaldoxime, and thiocyanate.
- Also, several inorganic ions in a solution absorb in the visible region, *e.g.*, salts of Ni, Co, Cu, *etc.*, and elements with incomplete inner electron shells.

2.4. Absorption of Solvents

- Solvents play a major role in biological spectroscopy. The spectrum of a compound in one solvent may be completely different in another solvent. Solvent polarity plays a significant role in determining the position and intensity of the absorption maximum of a compound.
- The solvents used for any spectroscopic measurement should ideally be transparent and not absorb the light wavelength used. Solvents that are commonly used in UV/Vis spectroscopy are water (wavelength cut-off 190 nm), acetonitrile (190 nm), cyclohexane (195 nm), methanol (205 nm), and ethanol (205 nm).
- Water, the biological solvent, is transparent to the UV/visible region of interest, *i.e.*, the wavelength above 190 nm.

3. Transmittance and Absorbance

- When light passes through a sample, the amount of light absorbed is the difference between the intensity of the incident radiation (I_0) and the intensity of the transmitted radiation (I).
- The amount of light absorbed is expressed as either transmittance or absorbance.
- Transmittance is usually given in terms of a fraction of 1 or as a percentage and is defined as:

$$T = \frac{I}{I_0}$$

or

$$\%T = \left(\frac{I}{I_0}\right) \times 100$$

- The absorbance can be calculated as follows:

$$A = -\log T$$

or

$$A = -\log\left(\frac{I}{I_0}\right)$$

or

$$A = \log\left(\frac{I_0}{I}\right)$$

- For most applications with UV/Vis spectroscopy, absorbance values are used since the relationship between absorbance and both concentration and path length is usually linear (discussed later in Beer Lambert's Law).
- **Figure 3** shows a simple example explaining absorbance and transmittance. In this example, 100 photons enter a cuvette and interact with the sample, and some photons are absorbed by the atoms to excite their electrons. Following absorbance, 20 photons are transmitted (leave out) and reach the detector. The transmittance and absorbance for this experiment can be calculated below:

Transmittance, $T = \frac{I}{I_0}$ $T = \frac{20}{100} = 0.2$

%Transmittance, $\%T = \left(\frac{I}{I_0}\right) \times 100$ $\%T = \left(\frac{20}{100}\right) \times 100 = 20\%$

Absorbance, $A = \log\left(\frac{I_0}{I}\right)$ $A = \log\left(\frac{100}{20}\right) = \log 5 = 0.669$

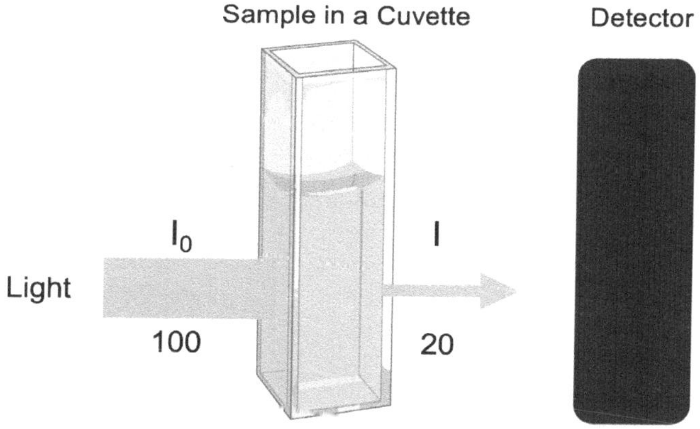

Figure 3: Absorbance and Transmittance Processes.
The details and calculations of the example are mentioned in the main text.

Note

If, T = 1.00 (100% T), A = 0.00

If, T = 0.10 (10% T), A = 1.00

If, T = 0.01 (1% T), A = 2.00

If, T = 0.001 (0.1% T), A = 3.00

- If the transmitted light is very less, the absorbance value will be high. Thus, very high absorbance values can suggest inaccurate data as the transmitted light falling on the detector is very minute.
- An absorbance value above 2.0 is always considered inaccurate data, as at high concentrations, the linearity of Beer-Lambert's law is lost.

4 Principle of UV/Vis Spectroscopy- Beer-Lambert's Law

- Beer-Lambert's law relates the changes in the light to the properties of the material through which the light is travelling.
- When monochromatic light passes through a solution in a cuvette, some of the light may be absorbed; thus, the intensity of the transmitted light is lesser than the incident light. The intensity of the emitted light depends upon the thickness of the cuvette and the concentration of the solution.
- The relation between the incident and transmitted light depends on the path length of the cuvette and the concentration of the absorbing solution. These factors are related to the laws of Lambert and Beer.
- The Beer-Lambert law states that the concentration of a substance in a solution is directly proportional to the absorbance, A, of the solution:

$$A \propto c$$

i.e., the intensity of transmitted and incident light remains related to the molar concentration of the compound and length of the solution:

$$A = \varepsilon c l$$

where,

A = absorbance (*no unit, we use AU- arbitrary unit*)

ε = *molar extinction coefficient or molar absorptivity,* ($L.mol^{-1}.cm^{-1}$)

l = path length of the cuvette (*cm*)

c = concentration of the compound in solution, ($mol.L^{-1}$)

Note

- The molar extinction coefficient (ε) is characteristic of a given substance.
- ε is a measure of how strongly a compound absorbs light at a particular wavelength.

- It is an intrinsic property of a compound that is dependent on its chemical composition and structure.
- It is calculated experimentally under a specific set of conditions.
- The molar extinction coefficient depends on the nature of the absorbing solute and the wavelength of the incident light used.

4.1. Limitations of the Beer-Lambert Law

- The linearity of the Beer-Lambert law is limited by several chemical and instrumental factors.
- The Beer-Lambert law is valid for low concentrations only. Higher concentrations lead to the association of molecules and therefore cause deviations from the ideal behaviour, and the linearity of Beer-Lambert's law is lost **(Figure 4)**. An absorbance value of above 2.0 is always considered inaccurate data. For practical purposes, absorbance values between 0.05 and 1.5 are considered reliable.

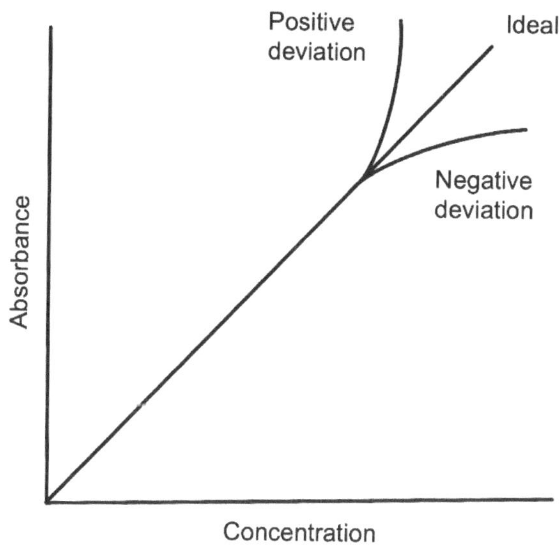

Figure 4: The Linearity of the Beer-Lambert Law is Lost at Higher Concentrations.

- The light may scatter if the sample contains particulates, leading to erroneous data.
- If the sample contains molecules that show fluorescence or phosphorescence, the Beer-Lambert law will not be followed.
- If the sample interacts with the solvent with hydrogen bonding, the data can have errors.

- ☆ The absorption spectrum of a chromophore is not only determined by its chemical structure but also by the solution environment. Hence the solution conditions (pH, polarity, redox state) may produce erroneous data.
- ☆ Deviations in ε can occur at high concentrations due to electrostatic interactions between nearby molecules.
- ☆ Changes in refractive index can happen at a high analyte concentration.
- ☆ Shifts in chemical equilibria can occur as a function of concentration.

5. Construction of a Spectrophotometer

- ☆ A spectrophotometer measures the amount of light a sample absorbs at different wavelengths.
- ☆ A spectrophotometer produces light of a pre-selected wavelength, directs it through the sample (usually dissolved in a solvent and placed in a cuvette), and measures the intensity of light transmitted by the sample.
- ☆ A light beam is passed through a sample, and the intensity of light reaching a detector is measured.
- ☆ **Figure 5** shows the four main components of a spectrophotometer: (1) Light source, (2) Monochromator (wavelength selector), (3) Sample cell holder, and (4) Detector.

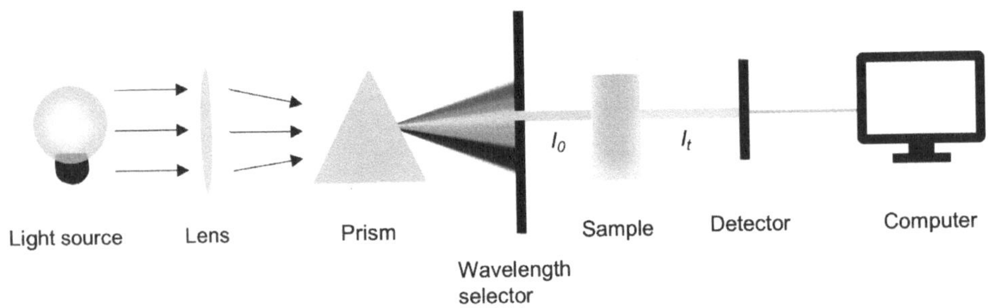

Figure 5: Schematic Representation of a Spectrophotometer.

5.1. Light Source

- ☆ The required properties of the light source of a spectrophotometer are:
  1. It should be a bright light source, *i.e.*, a high degree of uniform brightness across the range of wavelength measurements.
  2. The light source should be stable over time.
  3. It should have a low cost and long life.
- ☆ The spectrophotometers may have a deuterium lamp (for UV range) and a halogen or a tungsten lamp (for visible range).

- The spectrophotometers with a combination of tungsten/deuterium lamps are less efficient, as two lamps are required to cover the entire UV/Vis spectrum range. Moreover, both the lamps must be switched on during the entire instrument on time. This contributes to shorter lifetimes and higher instrument maintenance costs.
- Xenon lamps are more useful as they cover the entire UV/Vis spectrum range with only one lamp. In such a case, the lamp is kept only for the duration of the actual measurement. Thus, xenon lamps have a much higher lifetime (~10 years).
- The apparatus must be under a vacuum/inert environment while taking readings below 200 nm. At a low wavelength, oxygen absorbs light which can lead to the oxidation of deuterium lamps and the evaporation of tungsten at high temperatures.
- Some instruments may also use xenon or xenon flash lamps. They are used only for specific applications where very high-intensity light is required.

5.2. Monochromator

- A monochromator (or wavelength selector) creates the light of a single wavelength (monochromatic light) from the polychromatic light.
- It is a dispersion device that selects a particular wavelength of light from a wide range of light wavelengths generated from a source.
- The monochromator causes different wavelengths of light to be dispersed at different angles.
- The two commonly used types of monochromators are shown in **Figure 6**: (1) Prism, which splits light into a spectrum through refraction, and (2) Diffraction grating, which has a parallel serrated grating that diffracts light into a spectrum.

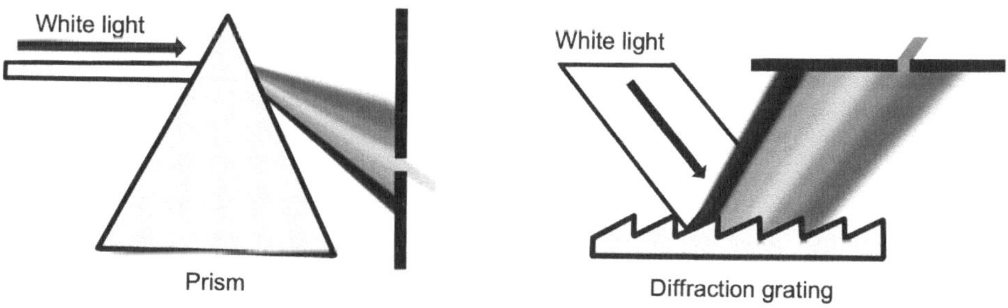

Figure 6: Two different Monochromators, *i.e.*, Prism and Diffraction grating, are shown.

- Before the monochromatic light hits a sample, it passes through a series of slits, lenses, filters, and mirrors. This optical system concentrates the light, increases the spectral purity, and focuses it toward the sample

- Spectrophotometers can have one or two monochromators. Most spectrophotometers are double-beam spectrophotometers (discussed in **section 6 and Figure 7**).

> ***Note***
>
> - Prisms are made of potassium bromide, sodium chloride, or caesium iodide.
> - Diffraction gratings are made up of alkali halides.
> - Filters are made up of lithium fluoride.

5.3. Sample Cell Holder

- Samples are kept in rectangular cells known as cuvettes, and the light beam is passed through the sample solution.
- A glass or plastic cuvette is used for collecting a spectrum in the visible range, while a quartz cuvette is used for the UV/Vis range (up to 190 nm).
- Oxygen absorbs light below 200 nm, so the apparatus must be under a vacuum/inert environment while taking readings below 200 nm.
- Nowadays, disposable plastic cuvettes are commercially available in polymethacrylate (280-800 nm) and polystyrene (350-800 nm).

> ***Note***
>
> - For accurate results, the sample to be analyzed must contain only the absorbing compound.
> - If the sample is a solution, a pure solvent sample should be used as a blank.

5.4. Detector

- A detector is a light-receiving element that absorbs light and converts it to an electrical response (signals), *i.e.*, it helps in photoelectric conversion.
- These signals can be detected. The resultant spectrum is a graph of the intensity of emitted or absorbed radiation vs wavelength. Signals provide quantitative and qualitative information on the sample being studied.
- Most detectors work on the principle of the photoelectric effect. Therefore, the current is directly proportional to the light intensity and its measure. The detectors generate electronic signals that are proportional to the transmitted light. These signals are translated into a readable form. This is done using amplifiers, ammeters, potentiometers, and potentiometric recorders.
- The most common detector is a photomultiplier tube. A photomultiplier tube detects a small amount of light, amplifies it by a cascade of electrons

accelerated by dynodes, and converts it into an electrical signal that can be fed into a meter or recorder.

- ☆ New detectors called photodiode arrays are also used in some modern spectrometers. They are composed of silicon crystals. Upon photon absorption by the diode, a current is generated that is proportional to the number of photons.
- ☆ The detector and amplifier are coupled to a computer. The computer stores all the data generated and produces the spectrum of the compound.

6. Types of Spectrophotometers

- ☆ Based on the light beam, the spectrophotometer can be of two types: Single-beam and double-beam spectrophotometer. Their components are shown in **Figure 7**.

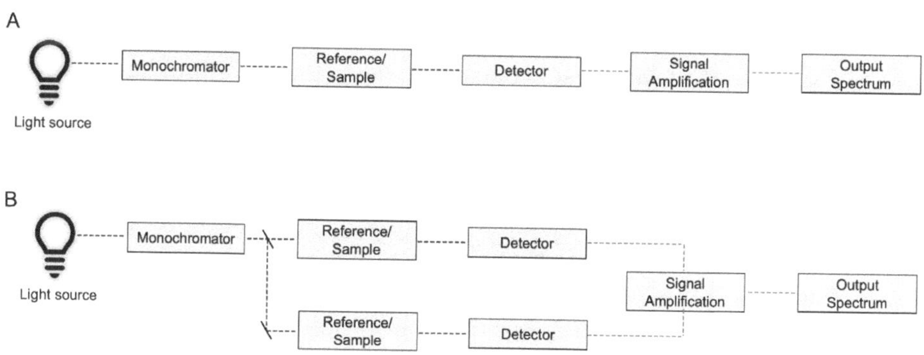

Figure 7: Types of Spectrophotometers.
(A) Single-beam spectrophotometer and (B) Double-beam spectrophotometer.

6.1. Single-beam Spectrophotometer

- ☆ In a single-beam spectrophotometer, a single light beam is used, which passes through the sample and reference cell one by one (*i.e.*, the test sample and the blank sample are measured one by one). The spectrum of the test and the blank samples are normalized manually to obtain the final spectrum. This can lead to significant errors over long time intervals in such spectrophotometers.

6.2 Double-beam Spectrophotometer

- ☆ In a double-beam spectrophotometer, two light beams are used, one passing through the sample while the other passing through a reference cell (*i.e.*, the test sample and the blank sample are measured together). The spectrum of the test and the blank samples are normalized automatically by the instrument, and the final spectrum is produced.

- It was developed to compensate for changes in lamp intensity between measurements on blank and sample cuvettes. The double-beam spectrophotometers provide more accurate and reproducible data.

7. Application of UV/Vis Spectroscopy

7.1. Determination of Molar Absorption Coefficient

- Using the Beer-Lambert law ($A = \varepsilon cl$), one can directly calculate the molar absorption coefficient of any compound.

7.2. Qualitative Analysis

- UV/Vis spectroscopy can be used to identify specific compounds.
- It provides limited qualitative information compared to other spectroscopic techniques (such as IR and fluorescence spectroscopy).
- It is useful in determining the structure of organic molecules, such as detecting the presence or absence of unsaturation, heteroatoms, functional group, *etc.*
- Most absorption by organic compounds results from the presence of π bonds or double bonds.
- When inserted into a saturated hydrocarbon (that exhibits no UV/Vis absorbance spectrum), π bonds may produce a compound that can show an absorption peak.
- For *e.g.*, The carbonyl ketone (C=O) group of acetone has a λ_{max} of 271 nm, and the carbonyl aldehyde (CHO) group of acetaldehyde has a λ_{max} of 293 nm, while the carboxyl (COOH) group of acetic acid has a λ_{max} of 204 nm.

> ***Note***
>
> - λ_{max}, the wavelength absorbance maximum, is the wavelength where the compound shows the highest absorbance.
> - However, the position of the absorbance maximum is not fixed. It depends on the environment of the chromophore and on the solvent in which the sample is dissolved. Other factors, such as pH and temperature, may also cause changes in both the intensity and the absorbance maxima.
> - The π bonds can show absorbance, but the intensity is very weak.
> - Conjugation of the multiple π bonds (greater extent of conjugation in a compound) increases both the intensity (hyperchromic shift) and the wavelength of the absorption maxima (bathochromic shift).

- The UV/Vis spectra do not enable absolute identification of an unknown compound. They are generally used to confirm the property of a compound through a comparison of the measured spectrum with a reference spectrum.

- UV/Vis spectroscopy can also be used to determine the physicochemical parameters of a compound and thus can provide information on the identity of a particular compound. For *e.g.*
  1. Proteins can be identified by their characteristic wavelength absorbance maximum at 280 nm due to the presence of aromatic amino acids (tryptophan, tyrosine, and phenylalanine).
  2. Nucleic acids can be identified by their characteristic wavelength absorbance maximum at 260 nm due to the purine and pyrimidine bases (A, G, C, T, U).
- UV/Vis spectroscopy can also be used to study enzyme activity measurements to evaluate its efficiency as a catalyst.

7.3. Quantitative Analysis

- UV/Vis spectroscopy is the most common method to determine the concentration of any compound.
- Various compounds like- amino acids (using ninhydrin test at 570 nm), proteins (using Folin's method at 660 nm; Biuret method at 540 nm; Bradford's method at 595 nm), co-enzymes (FAD at 438 nm; NADH at 340 nm; NAD^+ at 260 nm), nucleic acids (DNA, using diphenylamine at 595 nm; RNA using orcinol at 660 nm), *etc.* can be quantified using UV/Vis spectroscopy.

7.4. Enzyme Kinetic Assays

- Enzyme assays can be of two types: rate assays and endpoint assays.
- The rate of an enzyme depends on several factors, like temperature, pH, enzyme activity, and enzyme/substrate concentration. However, if other parameters are kept constant, the rate of reaction is directly proportional to the substrate concentration.
- With endpoint assays, the conditions are selected to complete substrate conversion to the product within a reasonable period (1–30 min). The difference between initial and final absorbance is directly proportional to the amount/concentration of the substrate.

7.5. As Detectors in Liquid Chromatography Instruments

- UV/Vis detectors are the most common detectors used in modern liquid chromatography systems.

7.6. Binding Studies

- UV/Vis spectroscopy can be used to understand the interaction between enzyme-substrate, enzyme-inhibitor, enzyme-cofactor, *etc.*

7.7. Monitoring the Growth of Microorganisms

- By monitoring the absorbance at 600 nm, the growth of microorganisms can be studied, and the growth kinetics can be studied.
- However, it should be noted that the underlying principle behind this is scattering and not absorbance. To differentiate scattering from actual absorbance, the observed value is referred to as the optical density (OD).

7.8. Determination of the Effect of Solvents

- A change in the polarity of the solvent changes the spectrum of a protein without changing its conformation. UV/Vis spectroscopy can be used to study the impact of the solvent on the protein conformation.

7.9. Determination of the Purity of Biomolecules

- The purity of macromolecules: DNA, RNA, or proteins can be checked using UV/Vis spectroscopy.
- The absorbance measurement of a nucleic acid solution at specific wavelengths is sufficient to determine its concentration and obtain insight into its purity.
- The quotients derived from the absorbance values measured at 260 nm, 280 nm, and 230 nm (A_{260}/A_{280} and A_{260}/A_{230}) constitute the purity ratios that help identify possible contaminations **(Table 2)**.

Table 2: Ratios Using the Absorbance at 260, 280, and 230 can Provide the Purity of DNA, RNA, and Protein.

The ratio limits presented are generally accepted ideal values.

Ratio	*Pure DNA*	*Pure RNA*	*Pure Protein*
A_{260}/A_{280}	1.8 >1.8 indicates RNA contamination <1.8 indicates protein contamination	2.0	0.57
A_{260}/A_{230}	2.3–2.4	2.1–2.3	-
A_{280}/A_{260}	-	-	2.0

- The A_{260}/A_{280} provides information on the type of nucleic acid present (DNA or RNA) as well as indicates its purity. Protein contamination results in a reduction in this ratio, while RNA contamination results in an increase in this ratio.
- Pure DNA shows an $A_{260/280}$ of 1.8; if the ratio is more than 1.8, it indicates RNA contamination, and if the ratio is less than 1.8, it indicates protein contamination.

- ☆ Pure RNA shows an A_{260}/A_{280} of 2.0; if the value deviates, it indicates contamination.
- ☆ Pure protein shows an $A_{280/260}$ of 2.0; if the value deviates, it indicates contamination.
- ☆ The A_{260}/A_{230} is an indicator of contaminants that absorb at 230 nm, such as chaotropic salts such as guanidine hydrochloride and non-ionic detergents such as Triton™ X-100 and Tween 20, EDTA, proteins, phenols, polysaccharides, *etc.*
- ☆ Pure DNA has an A_{260}/A_{230} ratio of 2.3–2.4, while pure RNA has an A_{260}/A_{230} ratio of 2.1–2.3.
- ☆ Thus, the A_{260}/A_{280} ratio of pure nucleic acid solutions is around 1.8–2.0, whereas the A_{260}/A_{230} ratio typically shows values around 2.0–2.5.
- ☆ Contamination with proteins affects the A_{260}/A_{230} more strongly than the A_{260}/A_{280}. Thus, the A_{260}/A_{230} is more suitable for determining protein contamination.
- ☆ A_{260}/A_{230} ratios generate a higher standard deviation than A_{260}/A_{280} ratios and, thus, should be interpreted with care.

Chapter 3

Fluorescence Spectroscopy

1. Introduction

- When atoms/molecules absorb UV/Vis light, their electrons can jump to higher atomic orbitals. Following this, the electron can return to its ground state by releasing the excess energy through collisions, transferring it to a nearby molecule, or emitting a photon (emission).
- Since the emitted photon carries less energy, it has a longer wavelength than the excitation photon. Thus, the emitted light can be distinguished from the excitation light.
- The emission of light following the absorption of a photon is also called photoluminescence.
- Photoluminescence can occur in two ways: fluorescence and phosphorescence (discussed later).
- Fluorescence is an emission phenomenon where an electron transition from a higher to a lower energy state is accompanied by the emission of radiation **(Figure 1).**
- In fluorescence spectroscopy, the energy released by the electron in the form of fluorescence is detected.
- The fluorescence spectrometer is also called the spectrofluorometer.

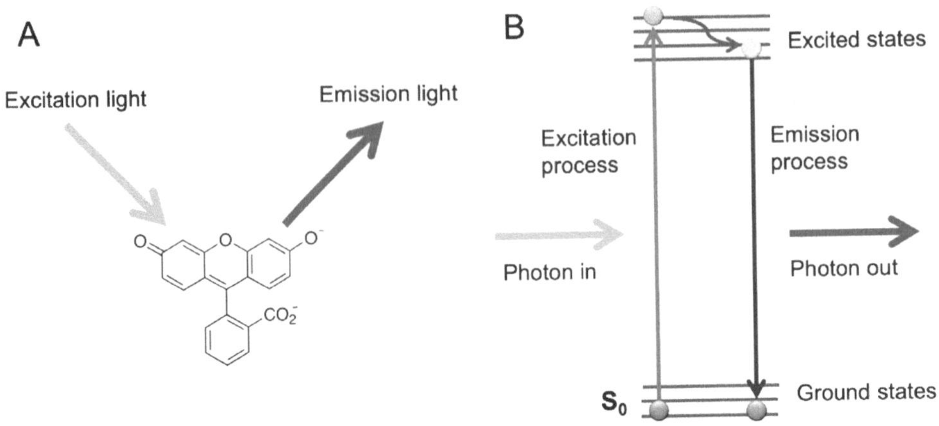

Figure 1: Basics of Electronic Transitions in Fluorescence Spectroscopy.

(A) A fluorophore can be excited with light, and in turn, it will emit light and return to the ground state. (B) A general schematic of the excitation-emission process.

> ***Note***
>
> - An atom/molecule may be excited in several ways, such as by absorbing a photon, thermal energy, or a chemical reaction.
> - Luminescence is the emission of light from any sample and occurs from electronically excited states.
> - When an atom/molecule is excited through a chemical reaction, the emission phenomenon is called chemiluminescence.

2. Principle of Fluorescence Spectroscopy

- Only molecules in their excited states can emit fluorescence; thus, they must be brought to a higher energy state before the emission phenomenon.
- Molecules have specific fluorescence properties that help determine, identify and characterize the molecule.
- Various electronic processes that occur following the excitation of an atom/molecule are shown using the Jablonski diagram **(Figure 2).**
- There are several processes and terms associated with the Jablonski diagram. They are discussed in the following sub-headings.

2.1. Absorbance

- Before excitation, the fluorophore (the fluorescing molecule, discussed later) is in the lowest energy state, *i.e.*, its electron is in the ground state (S_0), and the energy is minimum.

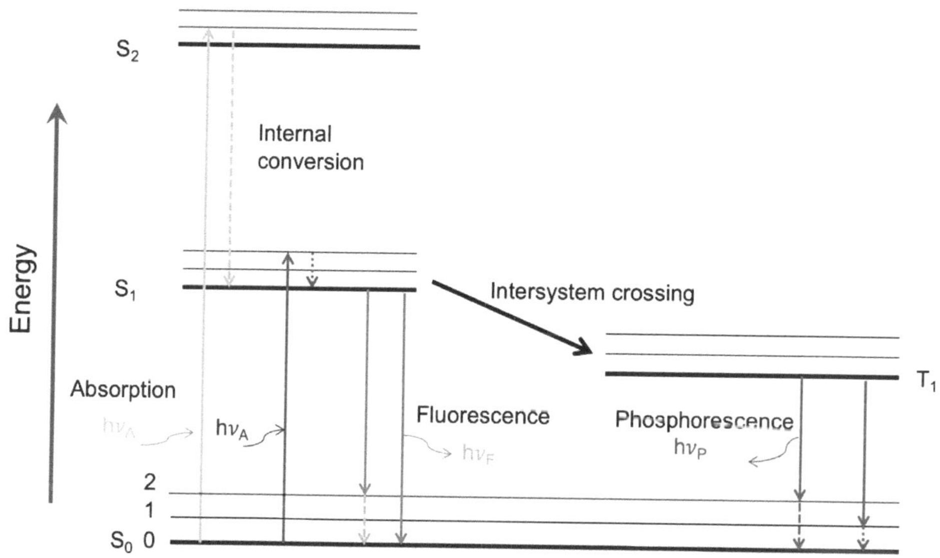

Figure 2: Jablonski Diagram Showing Various Processes following Light Absorption by a Fluorophore.

The S_0, S_1, and S_2 represent the singlet electronic states. The numbers 0, 1, and 2 illustrate the vibrational energy levels associated with the electronic states. T1 represents the first triplet electronic state.

- ☆ A fluorescent molecule can absorb light energy, and its electron can jump from a ground state orbital to an excited state orbital. This process is very fast (in the order of femtoseconds, *i.e.*, 10^{-15} seconds).

2.2. Internal Conversion

- ☆ The excited state electrons rapidly relax to the lowest vibrational state of S_1 through non-radiative processes, producing heat. This process is called internal conversion or vibrational relaxation.
- ☆ The timescale of internal conversion/vibrational relaxation is in the order of 10^{-12} seconds.

2.3. Fluorescence

- ☆ The electron in an excited state can lose energy and return to the ground state.
- ☆ Fluorescence is the emission of light resulting from the relaxation of the electron from the excited S_1 to ground S_0 states.
- ☆ The electron in the excited state orbital has an opposite spin as the ground state electron; therefore, the return to the ground state is spin-allowed.

- ☆ Since the return is spin-allowed, the excited state lifetime is short, *i.e.*, the rate is fast. The timescale is in the order of nano or picoseconds.
- ☆ The electrons return to the lowest vibrational state (S_0) through vibrational relaxation (from S_1 and S_2).

2.4. Intersystem Crossing

- ☆ It is possible that the electron in a vibrational state of S_1 can move to the isoenergetic vibrational state of T_1 through non-radiative processes. During this process, the spin of the electron is changed.
- ☆ Intersystem crossing refers to an isoenergetic non-radiative transition between electronic states of different spins.
- ☆ The electron then relaxes back to the lowest vibrational state of the triplet state (T_1).

2.5. Phosphorescence

- ☆ The electron in the T_1 state can lose energy and return to the ground state (S_0).
- ☆ Phosphorescence is the emission of light resulting from the relaxation of the electron from the excited T_1 to ground S_0 states.
- ☆ The electron in the T_1 state orbital has the same spin as the ground state electron; therefore, the return to the ground state is spin-forbidden.
- ☆ Since the return is spin-forbidden, the excited state lifetime is long, *i.e.*, the rate is slow. The timescale is in the order of microseconds or longer (even up to minutes).
- ☆ The electrons then return to the lowest vibrational state (S_0) through vibrational relaxation (from S_{01} and S_{02}), similar to fluorescence.

Note

- ☆ The fluorescence properties are determined by the properties of the molecule (internal factors) and the solution where the molecule is present (external factors).
- ☆ **Low fluorescence:** If the fluorophore emits a low fluorescence intensity, it might be challenging to detect, especially when background fluorescence is high. Fluorescence intensity can be enhanced by increasing the number of fluorescent molecules at the target site.

3. Important Terminologies in Fluorescence Spectroscopy

3.1. Fluorescence Intensity

- ☆ Fluorescence spectroscopy also follows Beer–Lambert's law, just like UV/Vis spectroscopy.

- The fluorescence intensity indicates the number of photons emitted. It is the extent of emission.
- Thus, the fluorescence intensity (height of the fluorescence peak) of a sample depends on the same parameters as absorbance, *i.e.*, F = εcl. Fluorescence intensity is the product of the molar extinction coefficient, optical path length, and sample concentration.
- However, the fluorescence intensity also depends on the fluorescence quantum yield of the molecule and the intensity of the excitation light.

3.2. Background Fluorescence

- The fluorescence detection by a fluorophore can be hampered by high background fluorescence. This can be due to the incomplete removal of unbound fluorophores or autofluorescence by the sample.

4. Effect of Solvents on Fluorescence Spectra

- Like absorbance spectroscopy, solvents also play a significant role in fluorescence spectroscopy.
- Fluorescence spectra are strongly dependent on solvent. The fluorescence spectrum of a fluorophore in one solvent may be completely different in another.
- Solvent plays a vital role in determining the position and intensity of the fluorescence spectra of fluorophores with large excited-state dipole moments. For such fluorophores, the fluorescence spectral shifts to longer wavelengths in polar solvents and vice versa.

5. Characteristics of the Fluorescence Phenomenon

- The entire fluorescence process is cyclical (repeated in cycles). Unless the fluorophore is irreversibly destroyed in the excited state (photobleaching phenomenon), the same fluorophore can be repeatedly excited and detected.
- The absorption and fluorescence spectrum of a hypothetical molecule is shown below **(Figure 3)**. We will use them for the discussions later in the chapter.

5.1. Stokes Shift

- A fluorescence emission spectrum of a sample is always shifted towards longer wavelengths with respect to the absorption spectrum. This shift is known as the Stokes shift.
- Therefore, the emitted radiation is of lower energy (*i.e.*, higher wavelength). This is expected as excited state electrons may lose energy through processes like internal conversion.

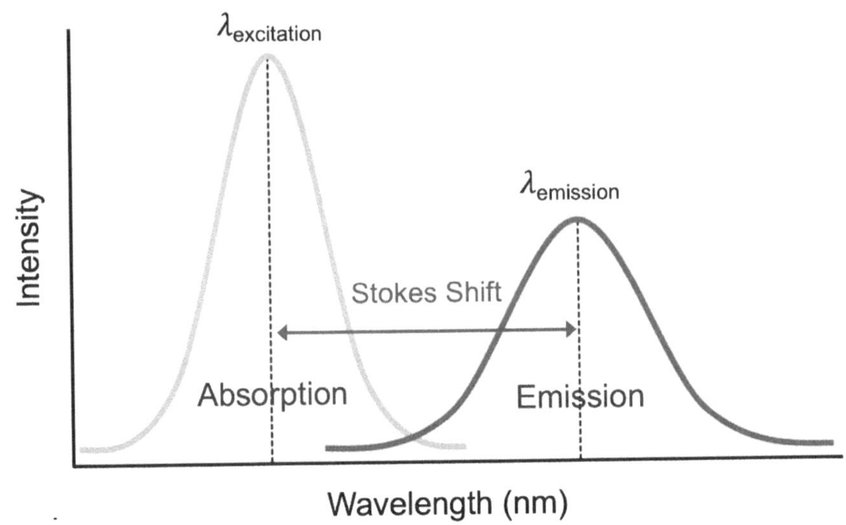

Figure 3: The Absorption and Emission Processes.

The absorption wavelength is lower than the emission wavelength (*i.e.*, the absorption energy is higher than the emission energy). The difference between the absorption and emission wavelength is called the Stokes shift.

- ☆ The difference in wavelength displayed by the absorbed and emitted photon is called the Stokes shift.
- ☆ Conversely, fluorophores with large Stokes shifts are easily distinguished because of the large separation between the excitation and emission wavelengths.
- ☆ Stokes rule = The emission wavelength is higher than the excitation wavelength.

5.2. Kasha's Rule

- ☆ Kasha's rule states that the fluorescence emission spectrum is independent of the excitation wavelength. This is due to the partial dissipation of excitation energy during the excited-state lifetime.
- ☆ The intensity of the emitted fluorescence light is proportional to the amplitude of the excitation spectrum at the excitation wavelength.
- ☆ In simple words, the greater the number of molecules absorbing a photon, the greater the number of molecules emitting a photon as fluorescence.

5.3. Franck-Condon Principle

- ☆ The Franck-Condon principle states that the atom nuclei are stationary during an electronic transition.

- The time taken for the electronic transitions (S_0 to S_1 states) is much faster than the time scale of nuclear vibrational motions. Thus, the nuclei can be considered unchanged during the electronic transition.
- This results in a relatively mirror-symmetric emission spectrum to the absorption spectrum.

5.4. Quantum Yield

- As mentioned earlier, an excited electron can return to the ground state through non-radiative pathways.
- The ratio of the number of photons emitted and absorbed by a fluorescent molecule (fluorophore) is called quantum yield (Q).

$$Q = \frac{\text{Number of photons emitted}}{\text{Number of photons absorbed}}$$

- The quantum yield denotes the efficiency of the fluorescence process. The higher the quantum yield, the brighter the emission.
- The maximum value of the quantum yield can be 1.0 or 100%; fluorophores with Q ~0.15 (15 %) and above are considered good.
- The quantum yield is a dimensionless quantity and, most importantly, the only absolute measure of the fluorescence of a molecule.
- Measuring the quantum yield is a complex process standalone. In biochemical applications, this measurement is rarely done.
- Most commonly, the fluorescence emissions of two or more related samples are compared, and their relative differences are analyzed.
- In water, the fluorescence quantum yields for tyrosine, tryptophan, and phenylalanine are 0.14, 0.13, and 0.024, respectively.

Note

- The brightness of a given fluorophore depends on the molar extinction coefficient (ε) and quantum yield (Q), both of which are specific for different fluorophores.
- The molar extinction coefficient is a measure of how strongly a compound absorbs light at a particular wavelength, while the quantum yield is calculated as the ratio of photons emitted and photons absorbed.
- The brightness of a fluorophore is calculated as the product of the molar extinction coefficient and quantum yield (ε x Q).

5.5. Fluorescence Lifetime

- A molecule spends a measurable amount of time in the S_1 state (in the range of 1-10 ns).

- The fluorescence lifetime is the time a molecule spends in the excited state before emitting a photon and returning to the ground state **(Figure 4)**.
- It is the time delay between the absorbance and the emission processes.
- The fluorescence lifetime is sensitive to the quantum yield of the molecule.

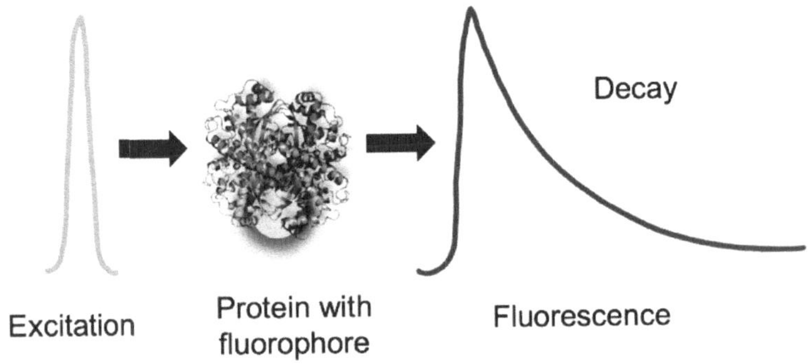

Figure 4: Lifetime Fluorescence Measurements.
Details are discussed in the main text.

6. Fluorophores

- Fluorophores (or fluorescent probes) are molecules or a part of molecules that absorb light of a specific wavelength and emit light of a longer wavelength (via fluorescence).
- A fluorophore is also known as fluorochrome, in analogy to a chromophore.
- Usually, fluorophores have several energy states, including those corresponding to different electron orbits and vibration/rotation states.
- Based on their location, fluorophores can be of two types: intrinsic and extrinsic fluorophores.

6.1. Intrinsic Fluorophores

- Intrinsic fluorophores are groups present within the molecule, *i.e.*, are naturally present in the sample to be measured.
- The aromatic amino acids tryptophan, tyrosine, and phenylalanine are intrinsic fluorophores in proteins. These fluorophores fluoresce in the near UV region (300-400 nm) **(Figure 5)**.
- Many cofactors and prosthetic groups, such as NADH, FAD, porphyrins, pyridoxamine, *etc.*, are also intrinsic fluorophores. These fluorophores fluoresce in the visible region (380-750 nm).
- In nucleic acids, the bases adenine (A), guanine (G), cytosine (C), and thymine (T) are intrinsic fluorophores. However, in contrast to proteins,

the intrinsic fluorescence from nucleic acids and individual nucleotides is very weak (due to this, we cannot see DNA directly under UV light).

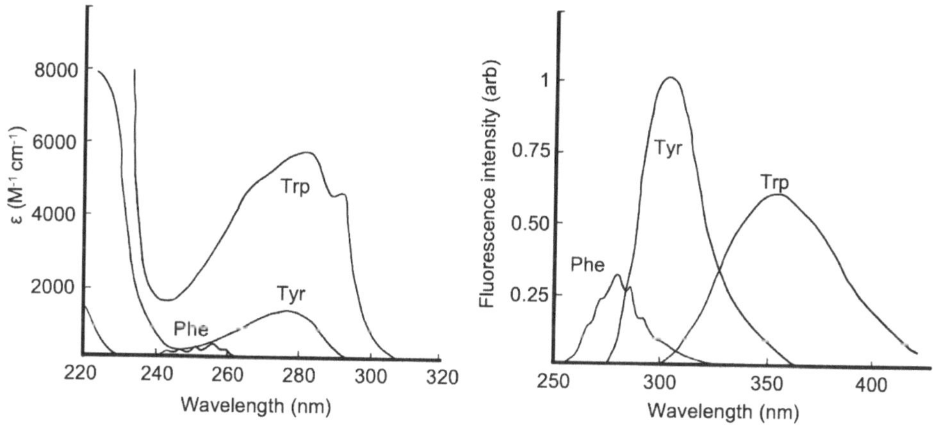

Figure 5: (A) Absorbance and (B) emission spectra of the aromatic amino acids. The absorption and fluorescence spectra of tryptophan, tyrosine, and phenylalanine are plotted as a function of wavelength.

6.1.1. Tryptophan Fluorescence

- The fluorescence of tryptophan is highly sensitive to its local environment. The emission spectra of tryptophan change in response to unfolding, subunit association, conformational transitions, or substrate/ligand binding.
- The fluorescence of protein is dominated by tryptophan.
- Tryptophan dominates protein fluorescence spectra as it has-
  1. high molar extinction coefficient (absorptivity),
  2. moderate quantum yield,
  3. ability to quench tyrosine and phenylalanine fluorescence emission by energy transfer (discussed later).

6.2. Extrinsic Fluorophores

- Extrinsic fluorophores are molecules that are attached to the sample, making them fluorescent. Examples include synthetic organic dyes, fluorescent proteins, and quantum dots.
- Dyes like ethidium bromide and acridine orange intercalate into double-stranded DNA. The intercalated dye increases the spacing of successive base pairs, distorts the regular phosphate backbone, and reduces the pitch of the helix **(Figure 6)**.

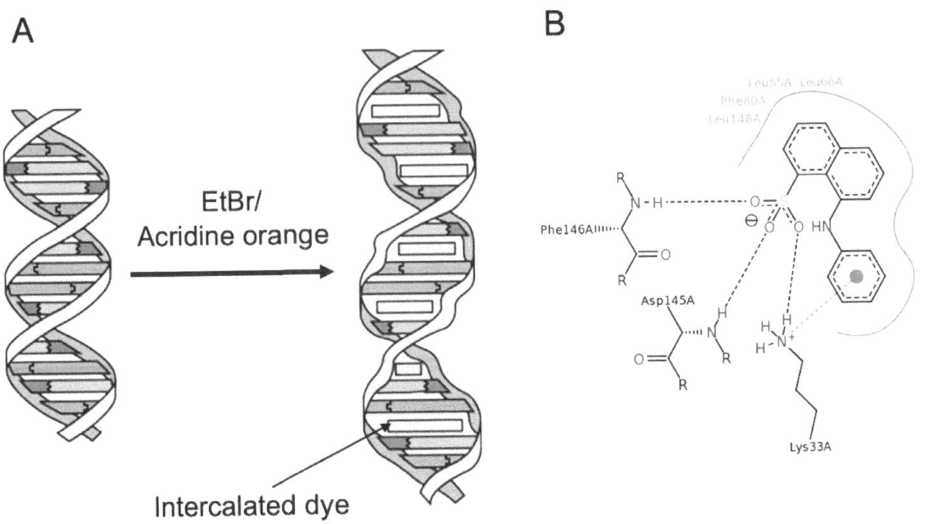

Figure 6: The Interaction of Extrinsic Fluorophores with DNA and Protein.

(A) The intercalation of ethidium bromide/acridine orange to the DNA. (B) Molecular interactions between anilino naphthalene sulfonic acid with the amino acids of a protein.

- ☆ Synthetic organic dyes such as ethidium bromide, acridine orange, anilino naphthalene sulfonic acid, dansyl chloride, fluorescein, and rhodamine are commonly used to study several biological processes **(Table 1)**.
- ☆ Fluorescent proteins, such as green fluorescent protein (GFP), are very helpful in monitoring various biological processes, particularly when combined with microscopy (discussed below in the box).
- ☆ Quantum dots are tiny particles or nanocrystals made up of a semiconducting material of 2-10 nm diameter (10-50 atoms). These nanocrystals can produce specific colours based on particle size. Quantum dots are more photostable than synthetic organic dyes and have relatively narrow emission bandwidths. They have a variety of nanotechnology and biomedical applications (like medical imaging and biosensors), *etc.*

Note

- ☆ The biological applications of fluorescence often utilize intrinsic protein fluorescence, as they display useful intrinsic fluorescence. Carbohydrates, lipids, and biomembranes are essentially non-fluorescent, while the intrinsic fluorescence of DNA is too weak to be useful.
- ☆ **Green fluorescent protein:** GFP isolated from the jellyfish, *Aequorea victoria*, which glows in the dark. It consists of 238 amino acid residues that fold into an 11-stranded β-barrel structure wherein an α-helix passes through the barrel. Once the protein folds, the residues

serine65-tyrasine66-glycine67 in the α-helix form the fluorophore structure (p-hydroxybenylideneimidazolinone).

- ☆ GFP is an excellent fluorophore with a molar absorption coefficient of ~30000 $M^{-1}cm^{-1}$ at 395 nm with a fluorescence quantum yield of 0.79.
- ☆ It has a major excitation peak at 395 nm and a minor at 475 nm. Its emission peak is at 504 nm.
- ☆ GFP can be tagged to a protein by expressing the fusing gene (GFP gene fused with the gene expressing the desired protein). The GFP then acts as a reporter for all the processes where the linked protein is involved.
- ☆ Several colour variants of GFP have been generated through modifications in the residues that constitute the fluorophore. The development of several GFP variants with varying excitation and emission characteristics (such as phycocyanin, phycoerythrin, phycobiliproteins, phycoerythrocyanin, *etc.*) has enabled the labelling of the proteins differentially. This has allowed easy monitoring of multiple biological processes using fluorescence microscopy.

Table 1: Common Synthetic Organic Dyes used to Study Biological Processes

Dye	*Use*	*Mechanism*	*Excitation Wavelength*	*Fluorescence Wavelength*
Ethidium bromide	Most common dye to detect nucleic acids in gels	Intercalates between the base pairs in the double helix	254 nm	Orange fluorescence (605 nm)
Acridine orange	Cell-permeable nucleic acid binding dye(has dual fluorescence)	Interacts with DNA and RNA by intercalation or electrostatic attractions	For dsDNA at 502 nm For ssDNA or RNA at 480 nm	Green fluorescence when bound to dsDNA (525 nm) Red fluorescence when bound to ssDNA or RNA (650 nm)
Anilino Naphthalene Sulfonic acid (ANS)	To monitor protein conformational changes, protein interactions, membrane proteins, and lipoproteins	An amphiphilic fluorescent probe. It binds to the hydrophobic regions or partly folded sites in a protein	Unbound dye at 380 nm Protein-bound dye at 380 nm	Low fluorescence at 545 nm Strong fluorescence at 470 nm

Dye	*Use*	*Mechanism*	*Excitation Wavelength*	*Fluorescence Wavelength*
Alexa fluor® dyes	As cell and tissue labels in fluorescence microscopy and cell biology studies; can be conjugated to primary/secondary antibodies to amplify signals	The mechanism of action varies for each probe in the series	Variable; the excitation spectra cover the entire visible spectrum range and extend to infrared	Variable the emission spectra cover the whole visible spectral range and extend to infrared

7. Construction of a Fluorescence Spectrometer

☆ The overall design of a fluorescence spectrometer is similar to the UV/Vis spectrophotometer, except that it has a right-angle geometry, *i.e.*, the emission detector is kept at a 90° angle from the direction of the excitation source **(Figure 7)**. This is made to avoid interference with the transmitted excitation light.

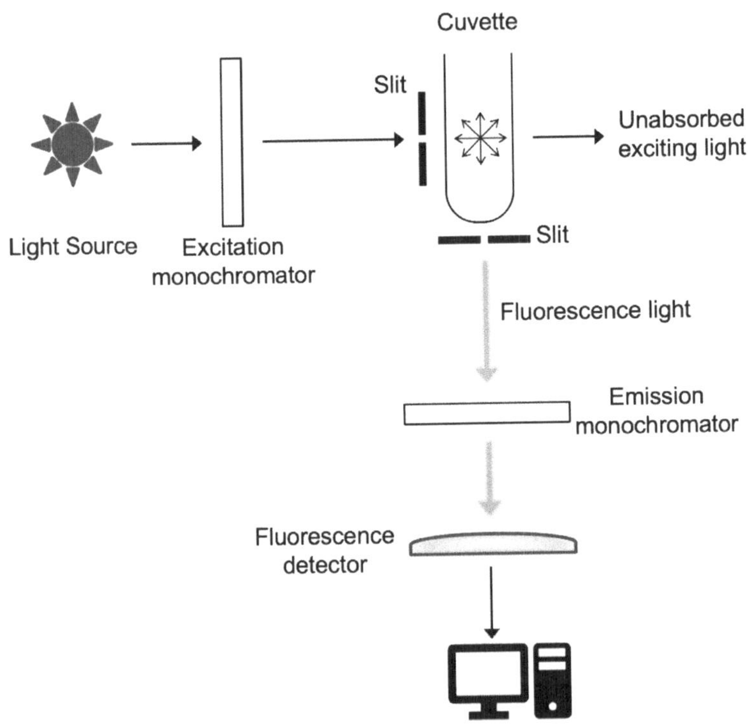

Figure 7: Construction of a Fluorescence Spectrometer.
The instrument has a right-angle geometry. Details are discussed in the main text.

☆ The main components of a fluorescence spectrometer are:

1. **Light source**- lasers, photodiodes, or lamps (xenon arcs and mercury-vapour lamps).
2. **Monochromator**- Diffraction grating. Two monochromators are used, one for tuning the wavelength of the exciting beam (excitation monochromator) and a second for analysis of the emitted fluorescence light (emission monochromator).
3. **Sample chamber-** quartz cuvette is typically used.
4. **Detector-** a photomultiplier tube to record the output, which is usually an electronic signal.
5. **Computer-** to collect and analyze data, sophisticated software packages enable high-end analyses of the spectra.

7.1. Types of Fluorescence-Based Instruments

☆ For practical purposes, there are four major types of fluorescence detection instruments:

1. **Fluorescence spectrometers/microplate readers** – record the average fluorescence properties in samples. For volumes in mL (fluorescence spectrometer) and samples in μL (microplate readers) are used.
2. **Fluorescence scanners/microarray readers** – detect localized fluorescence in samples in 2D for macroscopic objects (electrophoresis gels, blots, and chromatograms).
3. **Fluorescent microscopes** – detect localized fluorescence in samples in 2D or 3D for microscopic objects.
4. **Flow cytometers** – analyze the fluorescence of individual cells in a large population to be identified and quantitated.

☆ Other fluorescence detection instruments include DNA sequencers, capillary electrophoresis apparatus, and microfluidic devices.

8. Applications of Fluorescence Spectroscopy

8.1. To Study Protein Folding/Unfolding

☆ The tryptophan fluorescence depends on the localization of the indole side chain and the polarity of the solvent. It could shift due to conformational changes and protein unfolding, making it an excellent probe to study protein unfolding **(Figure 8)**.

☆ In proteins having tyrosine and phenylalanine, tryptophan can be selectively excited at 295 nm.

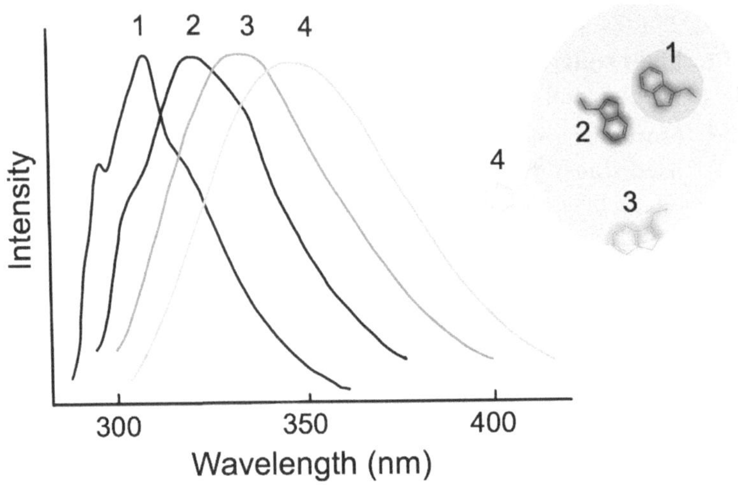

Figure 8: The Tryptophan Fluorescence Intensity Curves as a Function of Wavelength.

The inset shows the location of tryptophan in various positions in a globular protein. As the tryptophan environment shifts from the core to the outer side of the protein, there is a redshift towards the higher wavelength.

- In water and other aqueous solutions:
  1. When tryptophan is buried inside the protein core (in a hydrophobic environment), it emits fluorescence between 325-330 nm.
  2. When tryptophan is partly exposed, such as near the protein surface or intermediate states of protein folding, it emits fluorescence between 330-340 nm.
  3. When tryptophan is completely exposed to the solvent, as in the case of completely unfolded or denatured protein, it emits fluorescence between 340-355 nm.
- Thus, the unfolding of a protein can be studied by monitoring the shifts in tryptophan fluorescence as the buried tryptophan residues are exposed to the aqueous solution.
- An extrinsic chromophore, such as 1-anilino-8-naphthalene sulphonate (ANS), can be used to study the protein folding intermediates and molten globules. ANS emits weak fluorescence in a polar environment, *i.e.*, in an aqueous solution. However, in a non-polar environment, such as when bound to hydrophobic patches on proteins (partially folded intermediates), its fluorescence emission is significantly increased, and the spectrum shows a hypochromic shift; the emission maximum shifts from 545 nm to 470 nm. It is thus also a valuable tool for assessing the degree of non-polarity.

8.2. To Study Binding Kinetics

- ☆ The reaction rates and binding kinetics can be measured using the fluorescence kinetics method.
- ☆ To measure binding kinetics, the fluorescence spectrometer is coupled with a fast-mixing accessory called a stopped-flow instrument. The stopped-flow mixes two or more solutions within milliseconds.
- ☆ Fluorescence kinetics refers to the observation of fluorescence intensity over time.
- ☆ Fluorescence kinetics can be used to study the real-time binding of ANS to a protein to determine the binding rate.

8.3. To Study Binding Interactions

8.3.1. Using Fluorescence Anisotropy/Fluorescence Polarization

- ☆ The binding of a fluorophore to a biomolecule can be studied using fluorescence anisotropy or fluorescence polarization. It measures the rotational mobility of the fluorophores that are excited with polarized light.
- ☆ Measurement of steady-state and time-resolved fluorescence anisotropy can reveal important information on the molecular orientation, mobility, fluidity, rigidity, and processes that affect them. In addition, protein-protein interactions (crowding, oligomerization) and receptor-ligand binding can also be studied.
- ☆ Fluorescence anisotropy is a measurement of the change in the orientation of a molecule with the time between the absorption and emission events.
- ☆ The fluorophore binding to a protein reduces its tumbling (increasing its rotational correlation time), resulting in higher fluorescence anisotropy.
- ☆ Free fluorophores irradiated with plane-polarized light rotate quickly and emit depolarized light. Fluorophores bound to proteins tumble slowly and emit polarized light.
- ☆ The degree of molecular binding can be determined by measuring the amount of polarized emission, *i.e.*, by calculating the difference of the emission light intensity parallel (I||) and perpendicular (I) to the excitation light normalized by the total fluorescence emission intensity. The polarity of the emitted light depends on the protein size **(Figure 9)**.
- ☆ A ligand (substrate, cofactor, or inhibitor) binding to protein results in a change in polarization that can be detected.
- ☆ Studying the binding of a fluorophore to protein can reveal several important information based on the experiment designed and the type of fluorophore used.

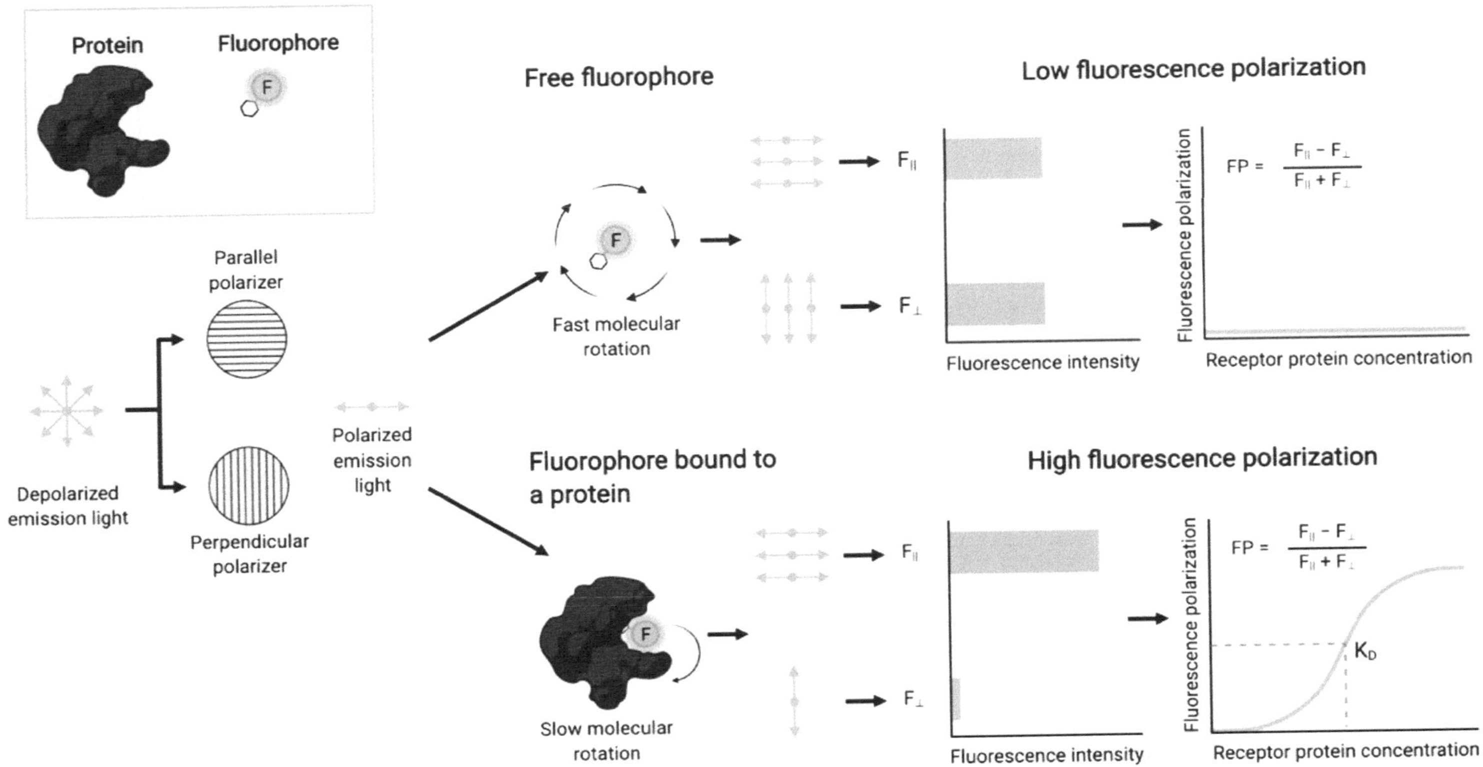

Figure 9: Principle of a Fluorescence Polarization Assay to Measure Protein–Ligand Interactions.

The fluorophore bound on the substrate is excited by linearly polarized light through a polarizer. Free fluorophores irradiated with plane-polarized light rotate fast, leading to the emission of depolarized light. Fluorophores bound to larger proteins rotate slowly and emit polarized light. The degree of binding can be determined by measuring the amount of polarized emission.

8.3.2. Using Fluorescence Quenching

- ☆ Fluorescence quenching refers to a decrease in fluorescence intensity.
- ☆ The emission of a fluorophore can be affected by the interaction with other fluorescent/non-fluorescent molecules that can quench the emitted fluorescence from the excited fluorophore.
- ☆ Fluorescence quenching can be defined as a process that reduces the fluorescence quantum yield without changing the fluorescence emission spectrum.
- ☆ Any molecule that can quench the fluorescence intensity of a fluorophore is called a quencher. Quenchers can be of two types: collision and contact quenchers **(Figure 10)**.

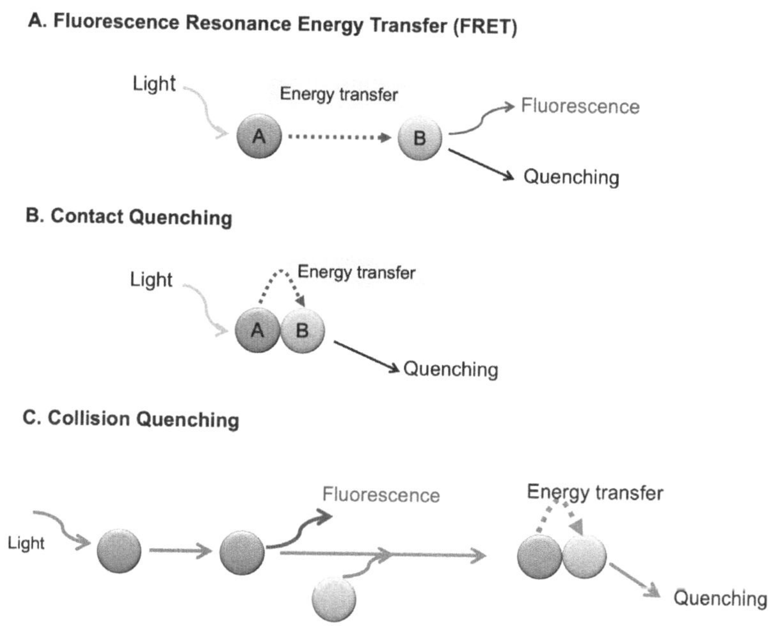

Figure 10: Fluorescence Quenching Processes. (A) FRET, (B) Contact quenching, and (C) Collision quenching. Details are discussed in the main text.

- ☆ Collision quenching occurs when an excited fluorophore interacts with a quencher molecule in an excited state resulting in the transfer of energy to the contact molecule and the relaxation of the excited fluorophore. A collisional quencher decreases the fluorescence emission intensity of a fluorophore by de-exciting the excited fluorophore through collisions. In other words, the addition of another non-radiative process to the system leads to a decreased quantum yield.
- ☆ Contact quenching occurs when a fluorophore interacts with a quencher in the ground state, resulting in the transfer of energy to the contact

molecule; this energy is then lost as heat. A contact quencher forms a non-fluorescent complex with the fluorophore. It decreases the effective concentration of the fluorophore resulting in a decrease in fluorescence emission intensity.

- Interaction of a fluorophore with another molecule may provide it protection against a collisional quencher.
- For *e.g.*, the interaction of a tryptophan-containing protein with a substrate/inhibitor can be studied using potassium iodide (KI) as a collisional quencher. A Stern-Volmer plot is generated to obtain meaningful data that can be predicted.
- The fluorescence intensity of a sample increases as the fluorophore concentration increases. However, beyond a certain concentration, the fluorescence intensity decreases due to self-collisional quenching. This property is explored to investigate several biological phenomena, such as membrane fusion, volume change, membranolytic activity, and membrane permeability. For *e.g.*,
    1. A fluorescent dye is suspended inside a lipid vesicle at self-quenching concentrations—the release of the fluorescent dye by the membranolytic molecule results in an increase in fluorescence emission intensity.
    2. The fusion of lipid vesicles can also be studied using the same approach. Vesicles that contain self-quenching concentrations of the fluorescent dye are titrated against the vesicles without fluorophores. A fusion will result in the dilution of fluorophores; the consequent decrease in self-quenching will result in an increase in the fluorescence intensity.

8.3.3. Using Fluorescence Resonance Energy Transfer

- Fluorescence resonance energy transfer (FRET) is a distance-dependent phenomenon in which the emission of one fluorophore is coupled to the excitation of the other.
- For effective FRET, the distance between the donor and acceptor molecule should be between 10–100 Å (Förster distance).
- Thus, FRET is also called Förster resonance energy transfer.
- In FRET, the excitation light raises a 'donor' molecule (fluorophore) to an excited state, which results in the release of a photon. An 'acceptor' molecule, which can be another fluorophore or a non-fluorescent molecule, is designed to be in close proximity to the donor fluorophore to absorb the emitted photon. The absorbance of the emitted photon by the acceptor fluorophore quenches the emitted light. If the absorbed photon is within the excitation wavelength range of the acceptor molecule,

the acceptor electron jumps to an excited state; consequently, it will release a photon at the emission wavelength of the acceptor fluorophore **(Figure 11)**.

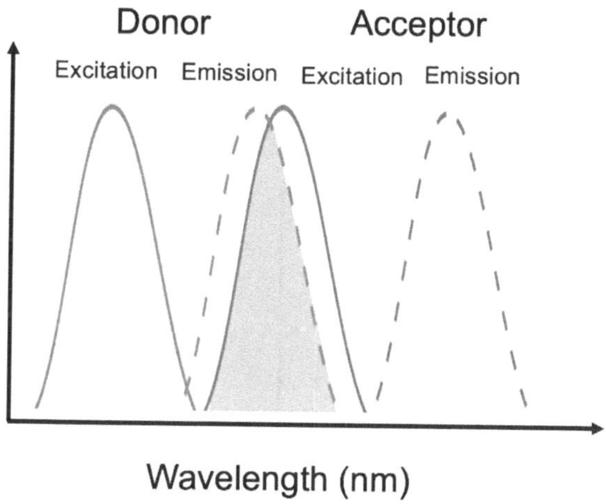

Figure 11: The Spectral Overlap in the FRET Process.
Details are discussed in the main text.

- The use of another fluorophore as a quencher is to detect the quenched state of the donor fluorophore because instead of detecting fluorescence at a known wavelength in response to excitation of the donor fluorophore, a quenched fluorophore would emit at the wavelength of the acceptor fluorophore.
- This non-radiative (without emission of a photon) transfer of excitation energy requires interaction between a donor and an acceptor molecule. This can occur if the donor's emission spectrum overlaps with the acceptor's absorption spectrum (*i.e.*, the vibrational and rotational transitions in the donor have the same energy as the corresponding transitions in the acceptor). Such transitions are coupled, *i.e.*, are in resonance **(Figure 12)**. Thus, the accurate term should be resonance energy transfer (RET) and not FRET because it is not the fluorescence that is transferred but the electronic energy of the donor that is transferred.
- FRET is a method to determine molecular proximity. Such estimation of inter-molecular distances in biomolecules has several applications.
- Biological phenomena that can be examined by FRET include protein-ligand interactions, protein-protein interactions, protein folding/unfolding, DNA denaturation/renaturation, protein-membrane interactions, membrane diffusion (both protein and lipid), *etc.*
- Fluorescence quenching and FRET are applications of fluorescence spectroscopy based on fluorophore–fluorophore interactions.

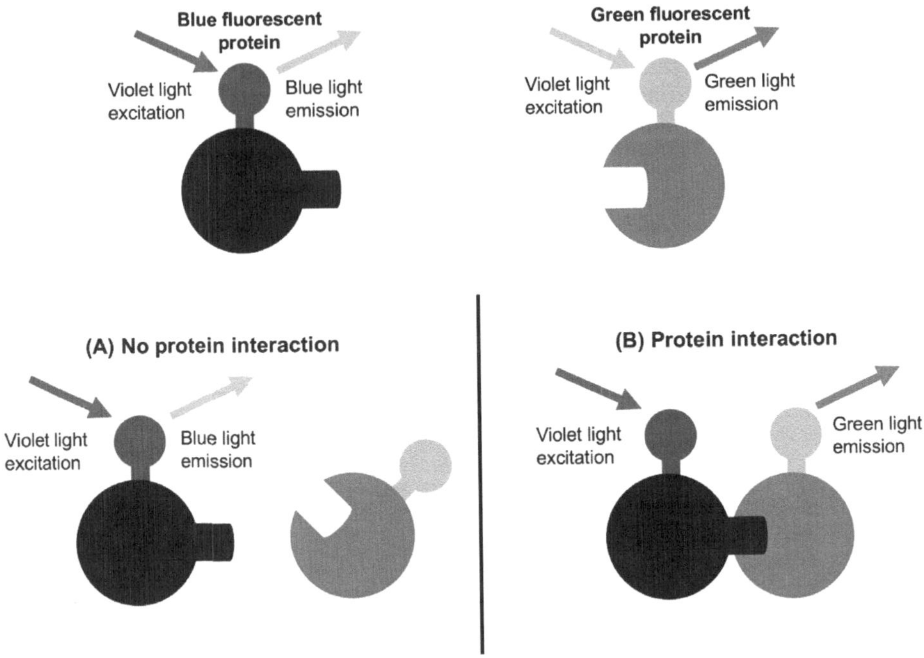

Figure 12: Fluorescence Resonance Energy Transfer.
Details are discussed in the main text.

8.4. Single-Molecule Fluorescence Spectroscopy

- Single-molecule fluorescence spectroscopy has emerged as a powerful tool to characterize the properties and interactions of biomolecules.
- Single-molecule methods have ultra-high sensitivity with enhanced throughput.
- Single-molecule fluorescence measurement is based on the repetitive excitation of a single fluorophore, generating repeated cycles of absorption and fluorescence and count rates of up to tens of thousands of counts per second.
- In a single-molecule spectroscopy instrument, laser light enters the fluorescence microscope from the back and is reflected by a dichroic mirror to the sample. The fluorescence is generally detected by single-photon counting modules.
- **Single-molecule fluorescence resonance energy transfer (smFRET)** – In contrast to normal FRET, which provides the FRET signal of a large number of molecules, smFRET can resolve the FRET signal of each molecule. It has been used to study protein folding and conformational dynamics and answer basic questions on replication, recombination, transcription, translation, RNA folding and catalysis, non-canonical DNA dynamics, *etc.*

- **Fluorescence correlation spectroscopy (FCS)** – In FCS, molecules are detected as they diffuse freely through the focal region of the fluorescence microscope. Fluctuations in fluorescence signals are analyzed by autocorrelation and cross-correlation of signals from one or more fluorescent species.
- **Single-pair FRET (spFRET)** – spFRET provides information on the distributions in resonance energy transfer of freely diffusing single molecules. It has emerged as a powerful method of studying conformational dynamics in proteins.
- Combining spFRET and FCS forms a powerful technique for understanding the conformational dynamics of proteins. Dynamics can be characterized in spFRET-FCS by cross-correlation analysis of signals from donor and acceptor fluorophores.
- **Single-molecule polarization modulation spectroscopy** – The polarization of light is used to excite a single molecule to probe the orientation dynamics of single molecules. As the polarization of the excitation beam is rotated, the absorption probability of an orientationally immobile molecule is changed, resulting in a high fluorescence modulation depth. In contrast, orientationally mobile molecules do not show modulated fluorescence.

Chapter 4

Circular Dichroism Spectroscopy

1. Introduction

- Circular dichroism (CD) spectroscopy is an important method for analyzing the structure and conformation of biomolecules.
- It measures the difference in the absorption of left and right circularly polarized light by optically active substances.
- To be CD active, a molecule must be structurally asymmetric and exhibit absorbance. Thus, only optically active (chiral) molecules can produce characteristic CD signals.
- Asymmetry can result from chiral molecules such as the peptide backbone of proteins, a non-chiral molecule covalently attached to a chiral molecule (aromatic amino acid side chains), or a non-chiral molecule in an asymmetric environment (*e.g.*, a chromophore bound to a protein).
- Most biomolecules are chiral and, thus, optically active.
- CD has vast applications in elucidating macromolecular structures, especially proteins and nucleic acids. For example, it is widely used to rapidly determine the secondary and tertiary structures of proteins, conformational changes, and thermodynamic parameters associated with the folding/unfolding of proteins.
- CD spectrometer is also called the spectropolarimeter.

2. Plane Polarized Light

- The properties of light are the basis for the working of the CD spectrometer.
- As discussed earlier, light is electromagnetic radiation where the electric and magnetic fields are always perpendicular to each other. From now on, we shall mention only the electric field, as the magnetic field vector is perpendicular to the electric field vector and the direction of light propagation at all points in time and space.
- Ordinary light (*i.e.*, non-polarized or unpolarized light) consists of several electromagnetic waves with their electric field vectors pointing in all possible directions. It means that the non-polarized light vibrates in all possible directions.
- In a plane polarized light, light is transformed from non-polarized light into a plane/linearly polarized light. That is, if the vectors in all but one direction are cut off, the resulting light is plane polarized **(Figure 1)**.

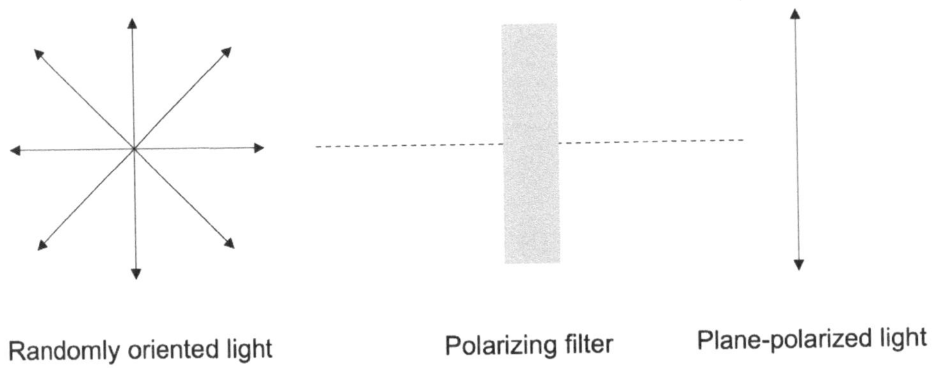

Figure 1: Generation of a Plane Polarized Light.

The plane polarized light is produced by passing the randomly polarized light through a polarizing filter or polarizer.

- In a plane polarized light, the electric field vector is in a single plane, *i.e.*, oscillating in a straight line.
- A plane polarized light can be produced by passing unpolarized light through a polarizer (photoelastic modulator).
- A polarizer is an optical fibre that passes only a specific polarization of light and blocks the rest polarization of light. It can be an anisotropic material (*e.g.*, polaroid) or a Nicol prism or grating **(Figure 1)**.

3. Circularly Polarized Light

- Plane polarized light consists of two circularly polarized components of equal intensity. The electric field vectors of the two circularly polarized components are like left- and right-handed springs **(Figure 2)**.

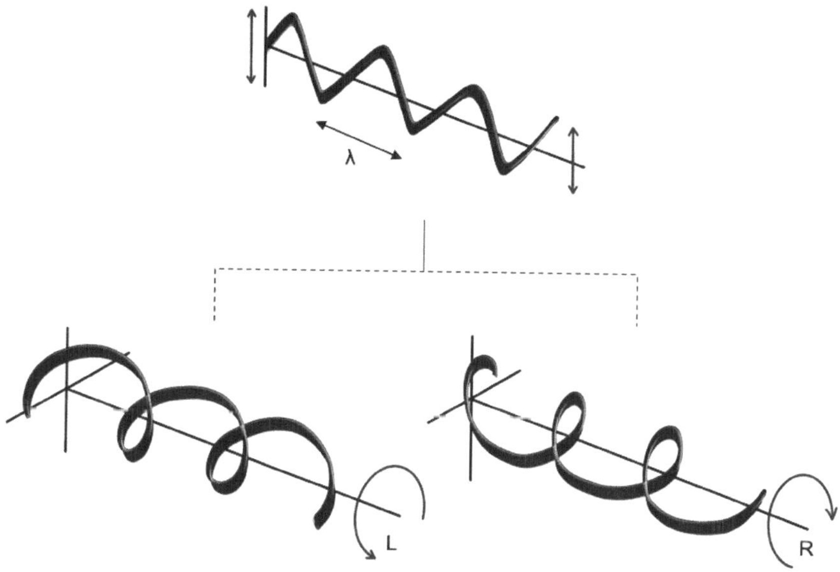

Figure 2: Components of a Plane Polarized Light.

A plane polarized light consists of two circularly polarized lights (left and right) of equal intensities.

- Thus, a plane polarized light can be viewed as a superposition of opposite circular polarized light of equal amplitude and phase.
- Looking from the source, the right-handed circularly polarized light rotates clockwise, while the left-handed circularly polarized light rotates anti-clockwise.

4. Principle of Circular Dichroism Spectroscopy

- Optically active molecules have different extinction coefficients for right (E_r) and left (E_l) circularly polarized lights.
- The circular dichroism (or ellipticity) arises from the differential absorption of the left and right circularly polarized light by an optically active compound **(Figure 3)**.
- When the left and right components are recombined after passing through an optically active medium, the result is an elliptically polarized light with ellipticity θ, *i.e.*, the plane polarized light becomes elliptically polarized **(Figure 3)**.
- Circular dichroism, therefore, refers to the differential absorption of the left and right circularly polarized light. It is defined as:

$$E = E_l - E_r$$

where, E_l and E_r are the absorbances for the left and right circularly polarized lights, respectively.

- Increased relative absorption of left polarized light results in a positive CD signal, while increased relative absorption of right polarized light results in a negative CD signal.

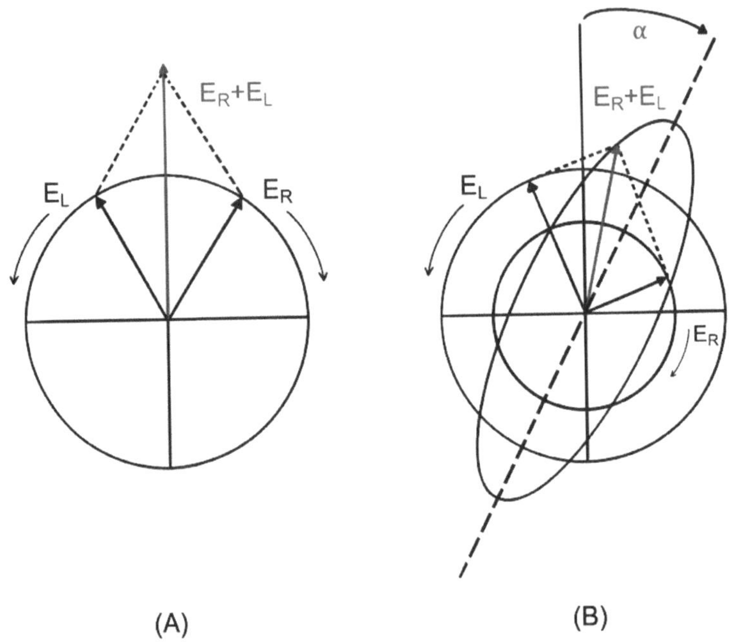

Figure 3: The Phenomenon of CD.

(A) Linear polarized light can be viewed as a superposition of opposite circular polarized light of equal amplitude and phase. (B) different absorption of the left- and right-hand polarized component leads to ellipticity (CD).

4.1. Units of CD Measurement

- CD data is represented in units of absorbance (Abs_l- Abs_r) or ellipticity (θ). Both of these can be normalized for the molar concentration of the sample. Absorbance is reported in the unit of milliabsorbance units (mA).
- CD is commonly reported as ellipticity (θ), which is related to absorbance by a factor of 32.98 ($\theta = 32.98\ \Delta Abs$). Ellipticity is reported in units of millidegrees (mdeg).
- For most calculations, the ellipticity θ is converted into a value called molar ellipticity ([θ]).

$$[\theta] = 100 \times \theta / c \times l$$

where, θ is the observed ellipticity in degrees, c is the concentration in molar, and l is the pathlength in cm. The use of the factor of 100 converts the pathlength into meters.

- ☆ In simple words, molar ellipticity ([θ]) is CD corrected for concentration. The unit of [θ] is $deg.cm^2.dmol^{-1}$.
- ☆ Conversion from molar extinction (absorbance corrected for concentration) to molar ellipticity ([θ]) uses a factor of 3298 ([θ] = 3298Δε). To calculate [θ], the sample concentration (g/L), cell path length (cm), and molecular weight (g/mol) must be known.
- ☆ If the sample is protein, the mean residual weight (average molecular weight of the amino acids it contains) is used instead of the molecular weight, essentially treating the protein as a solution of amino acids. In this case, the unit used is the mean residue ellipticity (MRE) or [θ]MR. This unit is specific to proteins. MRE is the normalized value of the molar ellipticity [θ], *i.e.*, it is a value independent of the length of a protein. MRE allows easy comparison of different proteins with different molecular weights.
- ☆ In simple words, MRE can be calculated by dividing the measured molar ellipticity [θ] of the protein by the number of amino acid residues in the protein. To calculate MRE, the concentration of protein (c, in molar) is multiplied by the number of amino acids (N) in the protein to calculate the mean residue concentration (CMR):

$$CMR = c \times N$$

- ☆ Then, the MRE or [θ]MR is calculated as:

$$MRE = 100 \times \theta/(CMR \times l)$$

- ☆ The unit of MRE is $deg.cm^2.dmol^{-1}residue^{-1}$.
- ☆ Thus, MRE is CD corrected for concentration and the number of residues in the protein.

Note

- ☆ The difference in absorption measured is very small, *i.e.*, the numerical value of the CD is tiny, and the actual effect is minute; hence the unit is millidegree (mdeg).
- ☆ Since the absorption difference is tiny, the detection must be as accurate as possible.
- ☆ Single-beam CD instruments require calibration with a CD standard. The most common standard used is (+)-10-camphorsulfonic acid.

5. Construction of a CD Spectrometer

☆ The main components of a CD spectrometer are shown in **Figure 4**. The main components are:

1. **Light source-** xenon lamps,
2. **Monochromator-** double prisms and a photoelastic modulator,
3. **Sample chamber-** quartz cuvette is always used.
4. **Detector-** a photomultiplier tube to record the output, which is usually an electronic signal.

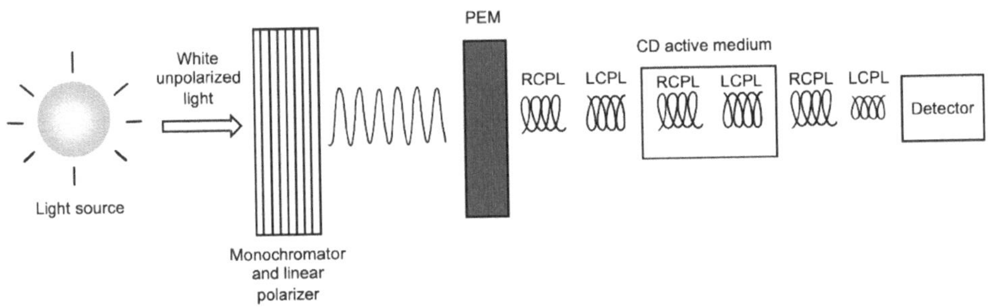

Figure 4: Schematic Diagram of a CD Spectrometer.
Details are discussed in the main text.

☆ Plane polarized light is produced by passing the light from the source (LS) through two prisms (P1 and P2) and a series of mirrors (M0 to M5), and slits (S1 to S3). Further, a photoelastic modulator (PEM) converts linearly polarized light into left- and right-circularly polarized light.

☆ The ordinary light (O) is focused by a lens (L) and passed through a filter (F) to the modulator (CDM).

☆ The circularly polarized light components are then passed through the shutter (SH) to the sample compartment and detected by the photomultiplier (PM).

☆ The instrument is kept under a constant purge of nitrogen as the xenon lamp creates a large amount of UV radiation that can interact with oxygen to form ozone. Ozone oxidizes the surface of the mirrors, decreasing their reflectivity and efficiency to focus light through the monochromator, reducing the signal-to-noise (S/N) ratio. Additionally, oxygen itself is absorbed in the UV region. By purging the monochromator with nitrogen, the mirrors retain their reflectivity, and the oxygen absorption is decreased, allowing the collection of data further into the far-UV region. Besides, the lifetime of the monochromator and mirrors is increased.

☆ An important consideration before collecting CD spectra is the selection of instrument parameters. The parameter settings such as scan rate, bandwidth, time constant, accumulation, and sensitivity should be adjusted per the study requirement.

6. CD of Biomolecules

- Biomacromolecules (proteins, nucleic acids, sugars, polysaccharides, glycerophospholipids, *etc.*) are intrinsically chiral due to their primary structure and presence of chromophores (such as the amide bonds and the aromatic side chains of proteins and the purine/pyrimidine bases of nucleic acids).
- Structurally, these molecules are intrinsically symmetric due to their planar geometry; however, once they fold into complex structures, the arrangement of these molecules in space becomes asymmetric, resulting in optical activity and giving rise to CD phenomena.

6.1. CD of Proteins

- CD spectroscopy can provide information about the protein structure, which may help understand its function.
- CD spectra contain information about the asymmetric features of the polypeptide backbone and the orientation of the side chains.
- Proteins possess several chromophores that can give rise to CD signals.
- Proteins show CD bands in two different spectral regions, the far-UV region (180–250 nm) due to the three common secondary structure motifs (α-helix, β-sheet, and random coil) and the near-UV region (250–300 nm) due to the aromatic amino acids.

6.1.1. Far-UV CD Spectrum of Proteins

- The chromophore that provides information about the conformation of the polypeptide backbone is the peptide bond; the spectra are therefore recorded in the far UV region, the region where the peptide bond absorbs.
- Each secondary structure component (α-helix, β-sheet, and random coil) has a specific CD spectrum **(Figure 5 and Table 1)**.

Table 1: The CD Signal for Various Secondary Structures and the Associated UV Transitions

Secondary Structure	*CD Bands*	
	Negative Dips	*Positive Peaks*
α-helices	222 nm (due to n $\rightarrow \pi^*$ transition) 208 nm (due to $\pi \rightarrow \pi^*$ transition)	190 nm (due to $\pi \rightarrow \pi^*$ transition)
β-sheets	216 nm (due to n $\rightarrow \pi^*$ transition)	198 nm (due to $\pi \rightarrow \pi^*$ transition)
Random coils or disordered structures	Below 200 nm (due to $\pi \rightarrow \pi^*$ transition)	218 nm (due to n $\rightarrow \pi^*$ transition), weak

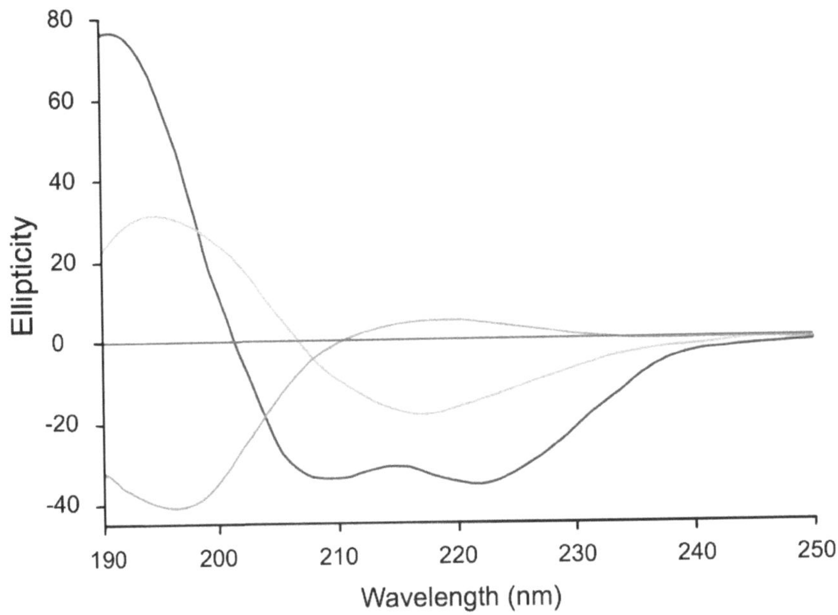

Figure 5: Far-UV CD Spectra for Pure Secondary Structures.

Representative far-UV CD spectra of α-helix (red), β-sheet (green), and random coil (blue) conformations are shown.

- ☆ The CD spectra of different secondary structures are distinct. Thus, it is possible to estimate the fraction of different secondary structures in a protein from its CD spectrum.
- ☆ The changes in the CD signal reflect changes in the protein structure.
- ☆ The principle chromophore that dominates in the far-UV CD spectrum of proteins is the peptide carbonyl bond (C=O) that absorbs at ~190 nm (due to $\pi \rightarrow \pi^*$ transition) and ~210-230 nm (due to $n \rightarrow \pi^*$ transition).
- ☆ The far-UV (180-250 nm) region detects the polypeptide backbone chain, whose phi and psi angles rotate depending on the protein conformation. These two absorption bands give rise to characteristic peaks/dips that can be used to calculate the protein's secondary structure components under different conditions **(Table 1)**.
- ☆ The CD curves and structural features of triosephosphate isomerase, lysozyme, myoglobin, and chymotrypsin are shown in **Figure 6**.

6.1.2. Near-UV CD Spectrum of Proteins

- ☆ The near-UV CD spectrum (250-320 nm) provides information on the tertiary structure of a protein.

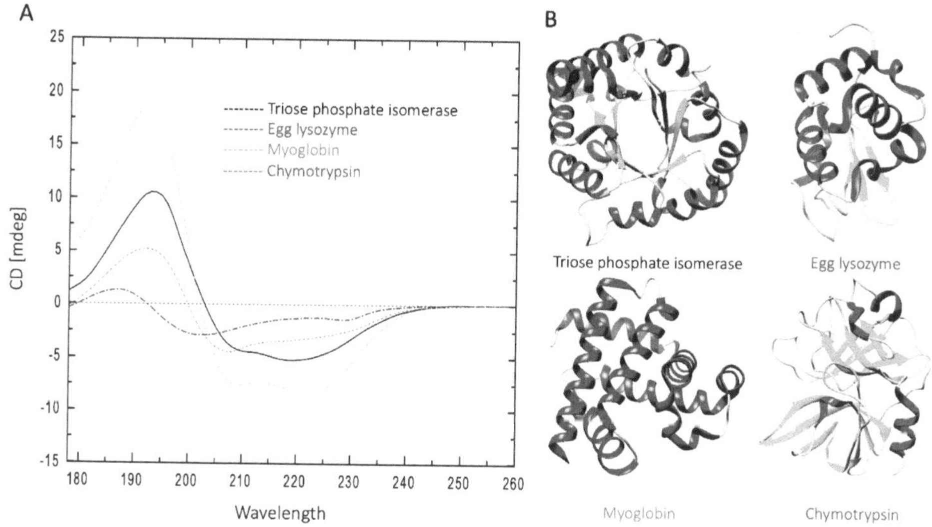

Figure 6: Structural Features of Proteins Triosephosphate Isomerase, Lysozyme, Myoglobin, and Chymotrypsin.

(A) Far-UV CD spectra of triosephosphate isomerase in black (helix 52 %, sheet 14 %, turn 11 %, other 23 %), hen egg lysozyme in red (helix 36 %, sheet 9 %, turn 32 %, other 23 %), myoglobin in green (helix 78 %, sheet 0 %, turn 12 %, other 10 %),and chymotrypsin in blue (helix 10 %, sheet 34 %, turn 20 %, other 36 %). (B) Cartoon representations of the crystal structures of the proteins.

- The three aromatic amino acid residues, phenylalanine, tyrosine, and tryptophan, show characteristic CD bands between 255-270 nm, 275-285 nm, and 290-305 nm, respectively **(Figure 7)**.
- The near-UV CD is a valuable fingerprint of protein conformation and orientation but does not provide detailed information on structural features.
- No standard spectra are available for near-UV CD; every protein is slightly different.

Note

- The CD spectrum is additive; the amplitude of the CD curve is a measure of the degree of asymmetry, *i.e.*, the total signal for a protein depends on its secondary structures.
- The strength of the CD signal is affected by several factors, such as i) the number and proximity of the aromatic residues, ii) the disulfide bonds, iii) the degree of hydrogen bonding, and iv) the rigidity of the chromophores in folded conformations.

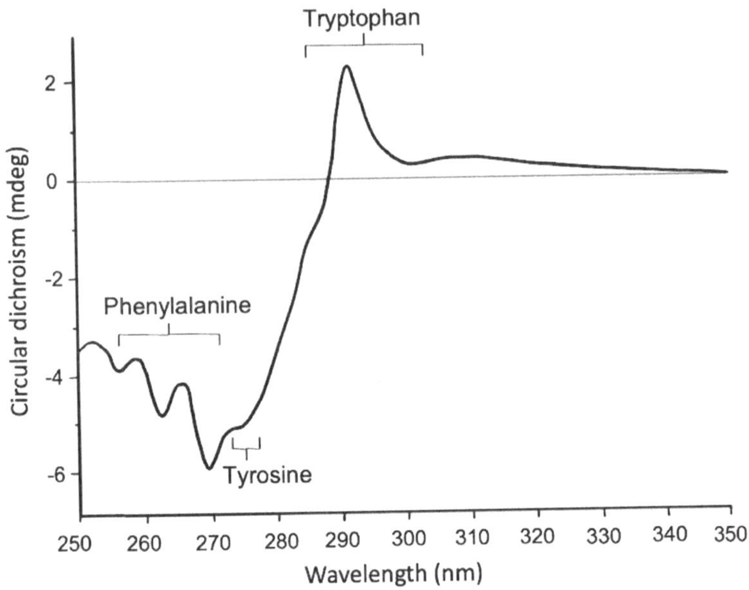

Figure 7: A Near-UV CD Spectrum.

A representative near-UV CD spectrum for a protein containing the aromatic amino acids phenylalanine, tyrosine, and tryptophan is shown.

Note

- ☆ Certain non-protein chromophores, such as FAD, NADP, and haem groups, can also give rise to CD signals, depending on the chromophore's precise environment.
- ☆ Many solvents and denaturing agents, such as guanidinium chloride and urea, absorb strongly below 200 nm. These solvents include water, which absorbs +90 % of the light below 182 nm, making CD acquisition in aqueous buffers impossible below ~185 nm.

6.2 CD Spectrum of Nucleic Acids

- ☆ The isolated purine and pyrimidine bases are not intrinsically chiral; they are optically inactive and hence do not exhibit a CD signal.
- ☆ However, when the base pairs are covalently bound to the DNA backbone, the glycosylic bond gives rise to a chiral structure. The chirality of the sugar groups induces a CD signal in the base pair's $\pi \rightarrow \pi^*$ transition, observed from 200-300 nm.
- ☆ The theoretical description of CD spectra DNA is very complex as the DNA molecules are very large. Hence CD spectroscopy cannot provide atomic-level structural information about the DNA molecules. For this reason, CD spectroscopy is primarily used empirically in studies of DNA.

☆ However, the DNA samples can be easily titrated with various agents to cause conformational isomerization, which can be studied using CD spectroscopy.

☆ The CD signal of nucleic acids increases with length due to the cooperativity of chiral interaction between contiguous bases.

☆ The CD spectra of nucleic acids in the UV region provide information on the secondary structure and conformational properties of the molecules. It can also provide information about the sequence composition of DNA from different organisms **(Figure 8)**.

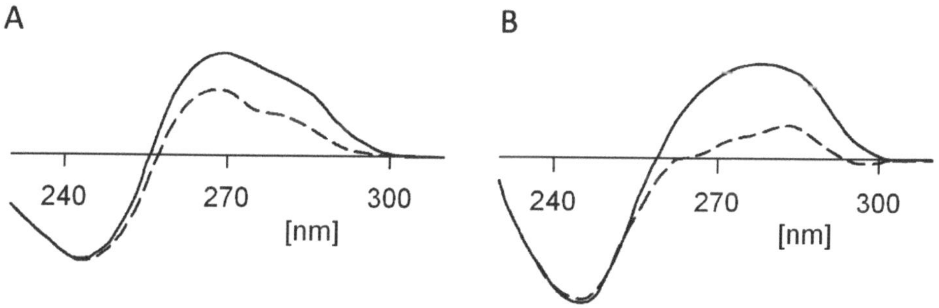

Figure 8: Sequence-Dependent CD Spectra of Native DNA of Two Organisms.
(A) *Sarcina lutea* (71 % G+C) and (B) Calf thymus (42 % G+C). Solid lines represent the spectra measured in 10 mM sodium acetate, pH 7, while the dashed lines represent the spectra measured in 5 M NaCl.

☆ Each nucleic acid conformation displays a distinct CD spectrum.

1. The B-form structure shows a positive peak at 290 nm, a negative dip around 245 nm, and a broad positive peak between 260-280 nm.
2. The A-form structure that is typically observed for RNA exhibits a positive peak at 260 nm and a negative dip at 210 nm.
3. The Z-form structure displays a negative dip at 290 nm, a positive peak at 260 nm, and a negative maximum at ~200 nm.
4. G-quadruplex (G4) structures show a positive peak at 290 nm and a negative dip at 260 nm (for anti-parallel structures) or a positive peak at 260 nm (for parallel structures).

7. Applications of CD spectroscopy

7.1. Determination of Secondary and Tertiary Structures

☆ CD spectroscopy is a quick and reliable method to study the secondary and tertiary structures of peptides, proteins, and nucleic acids, detect the percentages of secondary structures and monitor changes in the secondary structure over time.

- It is a useful method to compare the structures and conformations of proteins obtained from different sources or structures for various wild-type proteins with their mutants.
- CD spectroscopy is used to study the effects of the solution and solvents on protein secondary/tertiary structure. It is a widely used method to study conformational changes under the effect of salts, pH, temperature, and various solvents.

7.2. Study of Protein Folding/Unfolding and Stability

- CD spectroscopy is a common method to explore protein folding/unfolding pathways. The proteins can be denatured using chemical denaturants (such as urea or guanidine hydrochloride) or temperature or pH, and the folding/unfolding pathway can be determined. Also, the folding intermediates formed during the pathway can be identified and characterized.
- The thermodynamics associated with the folding can also be studied, and the various thermodynamic parameters (ΔG, T_m, ΔH, ΔS, $C_{1/2}$, m-value, ΔC_p, *etc.*) can be calculated.
- If the CD spectrometer is coupled with a stopped-flow measurement accessory, folding kinetics can also be studied.
- CD spectroscopy can also be used to study protein misfolding and aggregation.

7.3. Study of Protein Conformational Stability

- The conformational stability of proteins can be analyzed using CD spectroscopy.
- This can be achieved by denaturing the protein (as above) and detecting the structural changes using CD spectroscopy.
- In such studies, the CD signal is monitored at a single wavelength (such as 222, 208, or 216 nm) depending on the secondary structure under investigation. Plotting the change in ellipticity against increasing denaturant concentration/temperature provides the denaturation curve.
- The denaturation curves of three proteins (A- red, B- green, and C- blue) are shown below. Protein C is the most stable as it has the highest melting temperature (T_m), while protein A is the least stable as it has the lowest melting temperature **(Figure 9)**.

7.4. Study of Protein-Ligand Binding

- The binding of a ligand to a protein does not significantly affect the secondary structure of the protein; however, the local conformation of the tertiary structure might be changed if the ligand binds near the aromatic

residues. Such conformational changes can be detected using near-UV CD, and the binding can be studied.

- ☆ The binding of a ligand to a peptide can induce significant changes in the peptide's secondary structure. Such changes can be easily monitored using the far-UV CD.

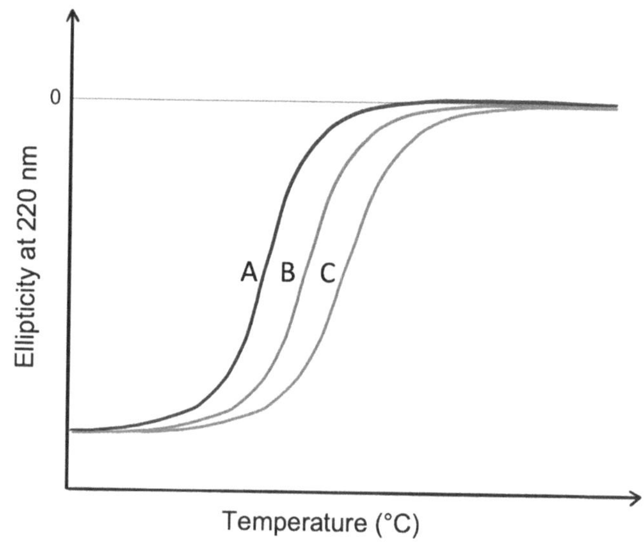

Figure 9: Denaturation Curves of Three Proteins, A, B, and C.
The order of thermal stability of the proteins is C > B > A.

7.5. Other Applications

- ☆ CD spectroscopy is particularly useful in establishing the structure-function relationship in proteins.
- ☆ The structural analysis of proteins that are difficult to crystallize (such as membrane proteins) can be performed using CD spectroscopy.
- ☆ CD spectroscopy is an effective method also to monitor protein-protein and protein-nucleic acid interactions.

Chapter 5

Infra-red Spectroscopy

1. Introduction

- ☆ We have discussed that all molecules absorb light of only a certain wavelength and transmit the rest.
- ☆ Photon energies associated with UV/Vis light are large. Thus, when molecules absorb light in the UV/Vis region, they undergo electronic transitions.
- ☆ Photon energies associated with infra-red (IR) are not large enough to excite electrons but may induce vibrations in covalently bonded atoms or groups. Detecting such vibrations is the fundamental basis of IR spectroscopy.
- ☆ When a molecule absorbs IR radiation, it undergoes vibrational or bending transitions while the molecule remains in its electronic ground state.
- ☆ Vibrational transition refers to the transition between different vibrational levels of the same electronic state.
- ☆ Since IR spectroscopy is used to probe molecular vibrations, it is also known as vibrational spectroscopy.
- ☆ Thus, if we measure where a molecule absorbs IR light, we can gain information about the nature of the chemical bonds in the molecule.
- ☆ IR spectroscopy is a valuable tool for identifying functional groups in organic molecules
- ☆ The wavelength of IR light ranges from 780–250000 nm (*i.e.*, 0.78-250 μm).

- In IR spectroscopy, the spectra are represented using wavenumber ($\bar{\nu}$).
- Wavenumber ($\bar{\nu}$) is the number of wavelength cycles per unit distance. Thus, it is a measure of frequency.
- A wavenumber of 100 indicates that there are 100 wavelengths per cm.
- Wavenumber ($\bar{\nu}$) in cm^{-1} is given by:

$$\bar{\nu}(cm^{-1}) = \frac{1}{\lambda(\mu m)} \times 10^4$$

- The IR region can be divided into three subregions: near-IR, mid-IR, and far-IR, as shown in **Figure 1**.
- The mid-IR region ($\bar{\nu}$ = 4000-400 cm^{-1}, λ = 2.5-25 μm) is the region that is used to study molecular vibrations and provides valuable information on the functional groups.

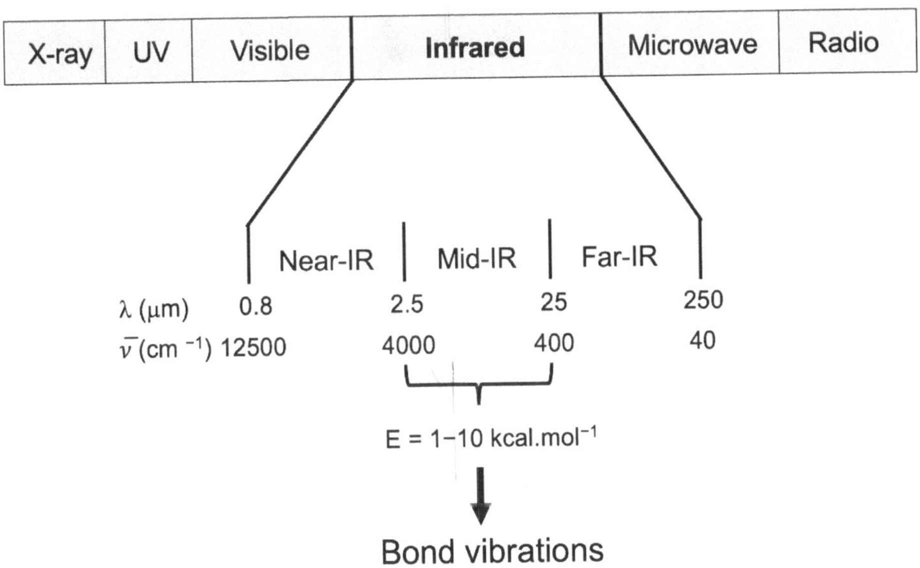

Figure 1: Infra-red (IR) Region of the Electromagnetic Spectrum.
The wavelength and wavenumber of the near-IR, mid-IR, and far-IR regions are shown.

2. Molecular Vibrations

- The covalent bonds in molecules are not rigid sticks or rods but are more like flexible springs that can be stretched and bent.
- Molecules have three types of motion: translation, rotational, and vibrational.

1. **Translation motion:** when molecules move from one position to another.
2. **Rotational motion:** entire molecule or parts of the molecule rotate with respect to one another.
3. **Vibrational motion**: when bonds between atoms within a molecule move.

- At non-zero temperatures, all the atoms in a molecule are in motion. The molecule itself is in translational and rotation motion.
- The bonds are under constant vibrational motion in different ways. A vibrating bond can be imagined as a spring with its ends connected to atoms (**Figure 2**).
- The molecular vibrations are very fast; the vibration cycles occur in a timescale of $\sim 10^{-15}$ seconds.

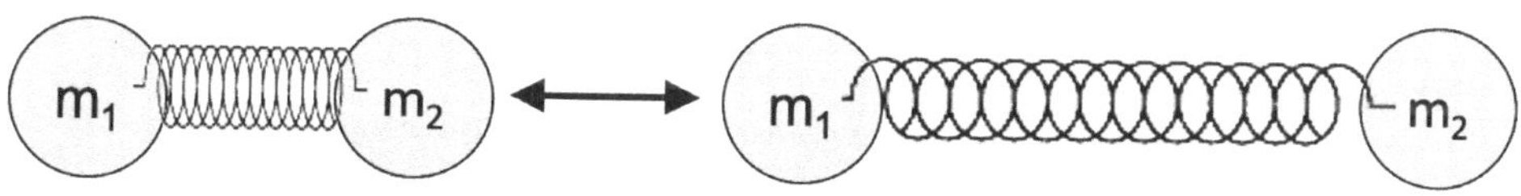

Figure 2: Spring Analogy of Bond Vibrations.

- An atom has 3 degrees of freedom (in the X, Y, and Z axes). Thus, a molecule consisting of N atoms has a total of 3N degrees of freedom.
- A non-linear molecule has a total of 3N-6 fundamental vibrations, and a linear molecule has 3N-5 fundamental vibrations.
- The vibrations can be either stretching or bending. Various types of stretching and bending vibrations are shown in **Table 1**, and their representation is in **Figure 3**.

Table 1: Different Types of Molecular Vibrations

Vibration type			
Stretching		*Bending*	
Symmetric	Asymmetric	In plane bending	Out of plane bending
		Scissoring, Rocking	Wagging, Twisting

- In stretching vibration, the bond length changes.
- In bending vibration, the bond angle changes.
- An IR spectrum originates when a molecule absorbs IR light of a particular wavelength, which is absent from the transmitted light. The resultant spectrum will show an absorption band.

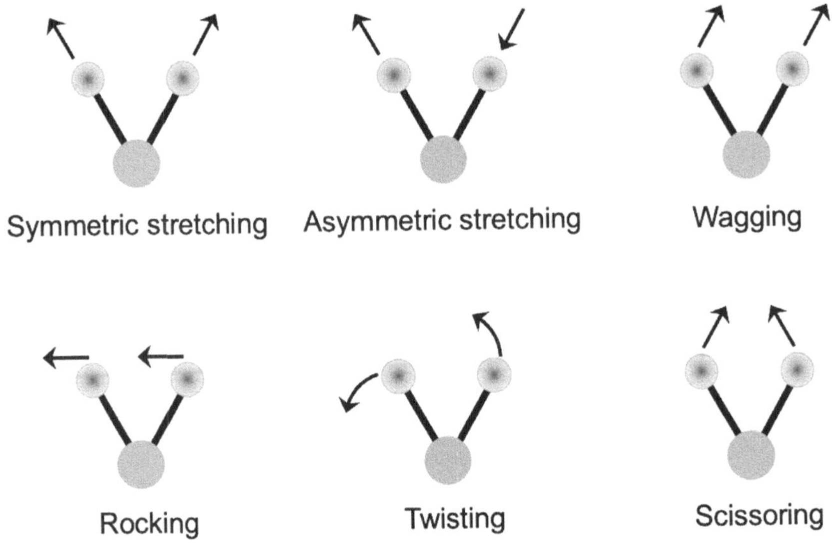

Figure 3: Different Types of Molecular Vibrations.

3. Absorption of IR Radiation

- A molecule will be IR active only if the vibrations give rise to a net change in the molecule's dipole moment. A few examples of IR active and non-active molecules are shown in **Figure 4**.

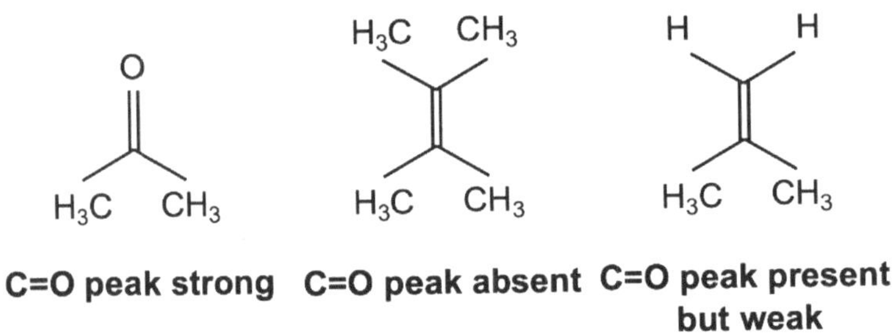

Figure 4: Examples of IR Active and Non-active Molecules.

Molecule 1 shows a strong C=O peak due to asymmetric stretching. Molecule 2 shows no C=O peak due to symmetric stretching. Molecule 3 shows a C=O peak but is weak.

- Asymmetric stretching and bending change the dipole moment of a molecule - such molecules are **IR active.**
- Symmetric stretching does not change the dipole moment of a molecule - such molecules are **not IR active.**
- Diatomic molecules (such as H_2, N_2, O_2, *etc.*) have one mode of vibration, and thus they do not absorb IR radiation as the vibration does not accompany a change in the dipole moment.

☆ However, vibration diatomic molecules with two atoms having large electronegativity differences are usually IR active. Thus, diatomic molecules such as carbon monoxide (C=O) are IR active.

Note

☆ The IR spectrum of a molecule depends on the polarity of the solvent.

☆ The IR frequency depends on the masses of attached atoms—the greater the mass of atoms, the lower the IR frequency at which the bond will absorb.

☆ Increasing the mass of one of the atoms leads to a decrease in the frequency of vibration.

☆ The frequency of vibration also depends on additional factors such as the delocalization of electrons, H-bonding, and substitutions at the nearby groups.

4. Characteristics of IR Spectra

☆ The result of an IR spectroscopic measurement is an IR spectrum. IR spectrum is a plot of measured intensity versus wavenumber (or frequency) of IR light.

☆ IR spectroscopy measures the vibrations of atoms; this can be used to identify the functional groups present in the molecule. Generally, stronger bonds and light atoms will vibrate at a high stretching frequency (wavenumber).

☆ The IR spectrum of water is shown in **Figure 5**. It is a simple IR spectrum.

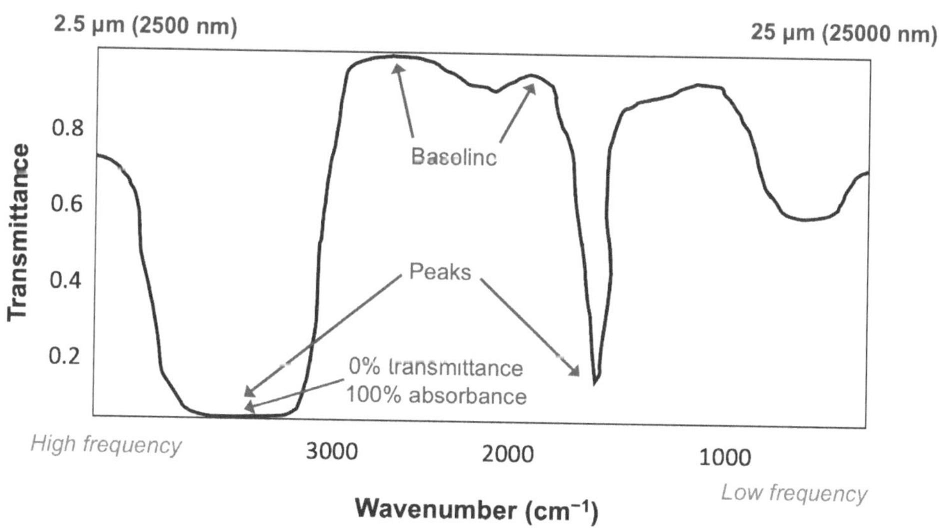

Figure 5: An IR Spectrum of Water. Details are discussed in the text.

- The region on the left (towards 4000 cm^{-1}) corresponds to high frequency and short wavelength; the region on the right (towards 400 cm^{-1}) corresponds to low frequency and long wavelength.
- The baseline at the top of the IR spectrum is not entirely flat.
- A water molecule shows major IR peaks at about 3200-3600 cm^{-1} (broad and strong), 1600 cm^{-1} (relatively sharp), and around 700 cm^{-1} (quite broad).
- The broad and weak peak around 2200 cm^{-1} is somewhat undefined. It is typically ignored.
- Note that each of these peaks corresponds to the vibration of a chemical bond in a molecule due to IR radiation. Hence, each molecule shows a characteristic IR spectrum corresponding to the bonds present in the molecule.
- An overview of the IR region (from 4000 cm^{-1} to 500 cm^{-1}) with the typical absorption values for various chemical bonds is shown in **Figure 6**.

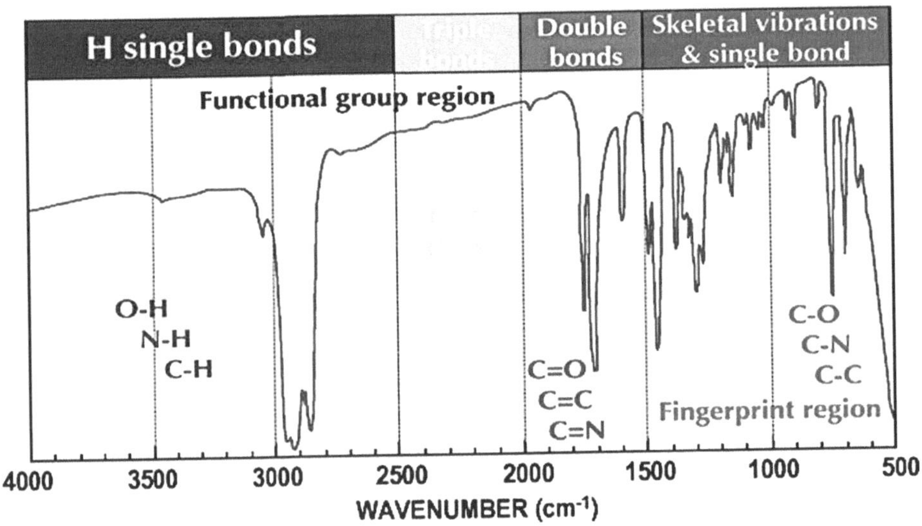

Figure 6: Typical IR Absorption Values for Various Chemical Bonds.

- The absorption bands in the mid-IR region between 4000–1500 cm^{-1} help identify functional groups. This region is therefore called the functional group region of the IR spectrum.
- The IR spectrum in the lower energy region between 1500–400 cm^{-1} contains a complicated set of peaks resulting from the complex vibrations involving several similar frequencies. This region is unique to a molecule and therefore is known as the fingerprint region of the IR spectrum. Though it is difficult to assign the vibrational modes to these peaks,

they are useful in the identification of a compound if the spectrum of the compound is already known.

5. Construction of an IR Spectrometer

- IR spectroscopy can be used to study solid, liquid, or gas samples.
- Commercially, two types of IR spectrometers are available: i) dispersive and ii) Fourier transform infra-red (FTIR) spectrometers.
- The main components of an IR spectrometer are shown in **Figure 7**.

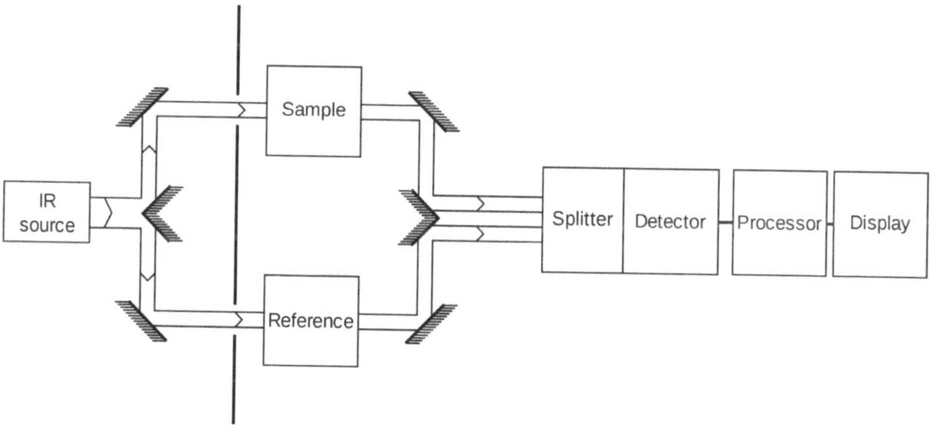

Figure 7: Components of an IR Spectrometer.

5.1. Dispersive Spectrometer

- A dispersive spectrometer is similar in design to a UV/Vis spectrophotometer. It has four major components: (1) a light source (inert solids that are electrically heated to radiate IR radiation), (2) a monochromator (reflective gratings and filters), (3) a sample cell holder, and (4) detector.
- The light source produces IR radiation. A light source is made up of inert solids that produce IR radiation upon heating electrically. Common light sources are Nernst glower, incandescent wire source, Globar, or a CO_2 laser. A combination of mercury arc and tungsten lamp is also used sometimes.
- The detectors are thermal detectors (such as a thermocouple or bolometer), photoconductive detectors (such as semiconductor materials), or pyroelectric detectors.
- The IR radiation generated by the source is dispersed into different frequencies by a monochromator. The selected frequencies pass through the sample and reference cells, and the transmitted light is measured by the detector. The detectors absorb IR radiation and permit conductance. Recorders are used to record the IR spectrum.

5.2. Fourier Transform Infra-red (FTIR) Spectrometer

- A Fourier transform spectrometer uses an interferometer in place of the monochromator. An interference of polychromatic radiation is generated using a Michelson interferometer. The absorption of IR light of a particular wavelength will bring a change in the interferogram that is detected. An interferogram is a time-domain signal and is converted to a frequency-domain signal through Fourier Transformation.
- Dispersive IR spectrometers are still in use, but FTIR has several advantages over dispersive ones.
- FTIR spectrometers are highly sensitive and accurate instruments that are very simple in design. The data can be obtained much faster.

6. Applications of IR Spectroscopy

6.1. Identification of Functional Groups in a Molecule

- IR spectroscopy is the most common method to identify the functional group present in a molecule. This is done by comparing the IR spectrum with library spectra using sophisticated software programs.

6.2. Identification of Compounds

- The fingerprint region of the IR spectrum is unique to a particular molecule. Thus, a molecule can be identified from its IR spectrum if the spectrum of the compound is already known and available for comparison.
- Since absorbance follows Beer Lambert's Law, it can be used for quantitative analysis.

6.3. To Study Protein/Peptide Structures

- IR spectroscopy is a routine method to study the structure and conformation of proteins and peptides. **Figure 8** shows the IR absorption spectrum of CRL1 protein and its second derivative in the amide I and amide II regions
- Like CD spectroscopy, the information from the peptide bonds is used to determine the conformation of the polypeptide backbone.
- The principal IR absorption bands used to characterize proteins are: amide I (1700–1600 cm^{-1}), amide II (1600– 1500 cm^{-1}), and amide III (1400–1200 cm^{-1}), which are due to the vibrational modes of the protein backbone.
  1. **Amide I (1700–1600 cm^{-1}):** This band is the most useful for studying polypeptide backbone conformation and secondary structures. The amide I band arises primarily due to the C=O stretching vibrations. The frequency of C=O stretching vibration is sensitive to H-bonding and, therefore, to the conformation of the polypeptide backbone.

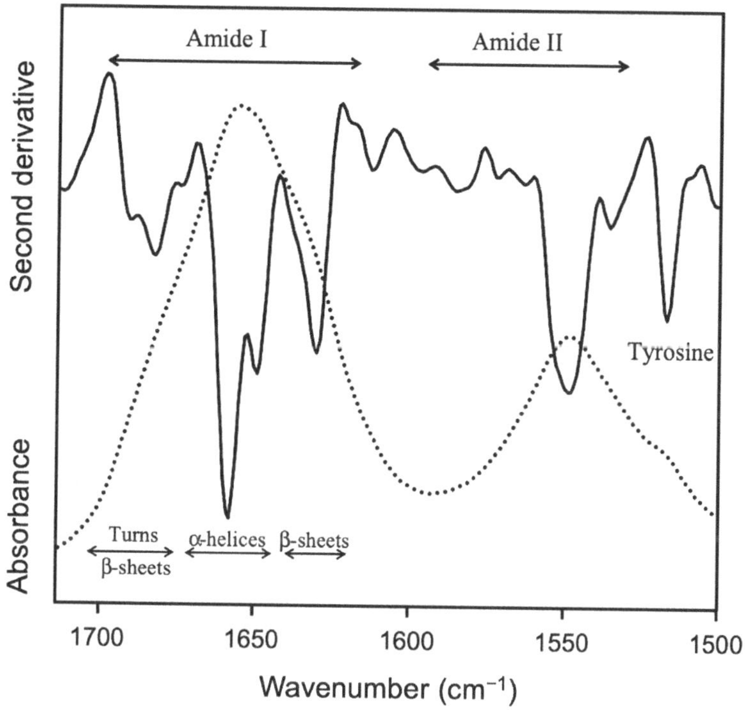

Figure 8: IR absorption Spectrum of CRL1 Protein in Water (Dotted line) and its Second Derivative in the Amide I and Amide II Regions (Continuous line). The tyrosine band around 1515 cm^{-1} is shown.

2. **Amide II (1600–1500 cm^{-1}):** This band arises due to the contribution of several backbone modes: NH in-plain bending and CN stretching, with minor contributions from C=O bending, CC, and NC stretching vibrations. The analysis of this band is complicated due to the overlapping of different vibrational modes. However, it is also sensitive to secondary structures.
3. **Amide III (1500–1400 cm^{-1}):** This band arises due to the contribution of several vibrations: the in-phase combination of NH bending and CN stretching, with minor contributions from C=O bending and CC stretching vibrations. This band has low intensity, and its analysis is highly complicated. However, this band is also sensitive to the backbone structures.

☆ The frequencies of absorption of different secondary structural elements are shown in **Table 2**.

☆ Similar to the CD spectrum, an IR spectrum of a protein can also be deconvoluted to determine the fractions of different secondary structural elements.

Table 2: The Absorption Frequencies of different Secondary Structural Elements in Proteins

Structure Type	*Wave Number (cm^{-1})*	
	Average	*Range*
α-helix	1654	1660–1648
Parallel β-sheet	1633	1640–1623
Antiparallel β-sheet	1686	1695–1675
β-turn	1672	1686–1662
Random coil/disordered	1654	1657–1642

6.4. Other Applications

- IR spectroscopy has emerged as a powerful tool to study protein conformation and dynamics. This can be done by identifying the protein absorption components in the amide I region (1700–1600 cm^{-1}) and assigning them to the secondary structural elements.
- IR spectroscopy is also used to determine the presence of impurities in a sample, structural transitions in lipids, nanoscale semiconductor analysis
- IR spectroscopy also allows studying the molecular mechanism of various diseases and helps identify specific spectral biomarkers for disease diagnosis.

Chapter 6

Raman Spectroscopy

1. Introduction

- Raman spectroscopy is a common method for studying molecular structures and is considered complementary to infra-red (IR) spectroscopy.
- Raman Spectroscopy is based on Raman Effect, discovered by Prof. C. V. Raman in 1928. For this discovery, he was awarded the Nobel Prize in Physics in 1930.
- Besides discovering the Raman Effect, Prof. Raman also studied X-ray diffraction, acoustics, optics, dielectrics, ultrasonics, photoelectricity, and colloidal particles.
- The Raman effect is based on the scattering of light, which includes both elastic (Rayleigh) scattering at the same wavelength as the incident light, and inelastic (Raman) scattering at different wavelengths due to molecular vibrations.
- Raman spectroscopy is a method based on the inelastic scattering of light. Inelastic scattering implies that the frequency of light photons changes upon interacting with the sample.
- Raman scattering is comparatively less intense than Rayleigh scattering. Therefore, to obtain Raman spectra, it is necessary to prevent Rayleigh scattering from overpowering the weaker Raman scattering.
- Photons are absorbed by the sample and then reemitted. As a result, the frequency of the reemitted photons is shifted up or down compared to the original frequency, which is called the Raman effect. This shift, called

the Raman shift, provides information about vibrational, rotational, and other low-frequency transitions in a molecule.

- ☆ When Light hits a sample, the atoms are excited and forced to vibrate. In Raman spectroscopy, these vibrations are measured.
- ☆ All types of samples, *i.e.*, solid, liquid, and gas, can be studied using Raman spectroscopy.
- ☆ Raman Spectroscopy is used to identify molecules and chemical bonds, protein structure and conformations, water interactions, *etc.*

2. Scattering of Light

- ☆ Light scattering is a physical process in which the light, upon interacting with molecules in a sample, is deflected in different directions **(Figure 1)**.

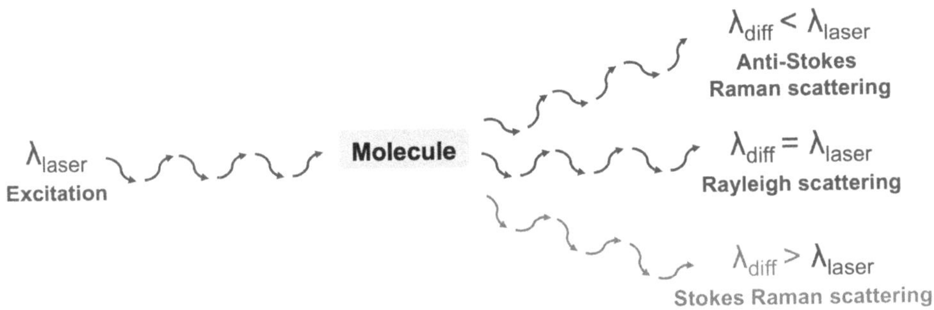

Figure 1: Scattering of Laser Light by a Molecule.

Stokes Raman scattering, Anti-Stokes, and Rayleigh scattering are shown. The difference in the frequency ($\nu=1/\lambda$) between the incident and scattered light can be used to characterize the molecule. Rayleigh scattering originates due to elastic scattering, while Stokes and Anti-Stokes (Raman) scatterings originate due to inelastic light scattering.

- ☆ Scattering means that the light photons deviate from their straight path after interacting with the sample.
- ☆ Thus, molecules can also be studied by examining the light scattered (In contrast to the transmitted light used in other spectroscopic methods).
- ☆ A laser is used as the incident light. In the experimental setup, the light detection systems are placed at specific angles to the incident direction, depending on the physical effect employed.
- ☆ The light scattering process can be either elastic or Inelastic.
- ☆ **Elastic scattering** happens when there is no loss of energy of the electron. Elastically scattered electrons may change direction but do not change their wavelength.

- **Inelastic scattering** happens when an interaction causes a loss of energy of the electron. Thus, inelastically scattered electrons have a longer wavelength (lower energy).
- The light scattering depends on: i) sample thickness, ii) interaction volume, and iii) electron transparency.
- There are many regimes of particle scattering,
- Several forms of light scattering can happen depending on the particle size, the light wavelength, and the refractive index. These include Rayleigh scattering, Debye or Mie scattering, Brillouin scattering, Raman scattering, Thompson scattering, *etc.*
- Raman spectroscopy relies on Raman scattering, while nephelometry and turbidimetry depend on Rayleigh scattering (discussed in chapter 11).

2.1. Rayleigh Scattering

- Rayleigh scattering occurs when the light interacts with particles smaller than the wavelength of light (such as atoms or molecules).
- It is a type of elastic scattering that results in scattered radiation in all directions uniformly.
- It depends on the wavelength, with shorter wavelengths being more scattered.
- Rayleigh scattering occurs in a dilute solution, enough for the particles to be considered independent scatterers.
- The blue colour of the sky we see is due to Rayleigh's scattering. The blue light from the sun is scattered much more than the red light in the upper atmosphere. Thus, the blue light is scattered overhead into the human eye while the red light remains unscattered and back into space.

2.2. Debye or Mie Scattering

- It is an elastic scattering that occurs from relatively large particles/ molecules with dimensions comparable to the incident light wavelength.
- The resulting scattered radiation is non-uniform, and the effect does not depend on the wavelength.
- The scattered white light seen in clouds or fog is due to Debye or Mie scattering.

2.3. Brillouin Scattering

- It is an inelastic scattering that occurs in light scattering from solid materials. The incident wavelength is modified by the energy levels of sound waves in the solid material resulting in minor shifts.

2.4. Thompson Scattering

- ☆ It is an elastic scattering where light is scattered by charged particles.
- ☆ A similar scattering called Compton scattering is an inelastic form of Thompson scattering that occurs when the energy of the incident light becomes comparable to the rest energy of the charged particle. Compton scattering is the primary attenuation mechanism of X-rays that provides contrast in medical X-ray images.

2.5. Raman Scattering

- ☆ It is an inelastic scattering of light where the frequency of the scattered light is changed by the gain or loss of energy that corresponds to energy levels in an atom or molecule.
- ☆ Raman scattering is very weak and is much smaller than the Rayleigh scattered light.
- ☆ This process is used for several diagnostic analyses.

3. Origin of Raman Spectra

- ☆ The Raman effect is based on changes in the electric field vector (E) of the incident light. This is determined by molecular polarizability (α). Polarizability can be understood as the ease of distorting a chemical bond.
- ☆ A peak will appear in the Raman spectrum only if there is a change in the polarizability of the molecule during the transition.
- ☆ For Raman-active vibrations, the incident light does not change the molecule's dipole moment but changes the polarizability.
- ☆ Upon interaction with the molecule, the laser light induces an electric dipole moment ($P = \alpha E$) that deforms the molecule. Because of periodical deformation, the molecule starts vibrating with characteristic frequency νm.
- ☆ The amplitude of vibration is called nuclear displacement.
- ☆ Thus, monochromatic laser light with frequency ν_0 can excite a molecule and transforms it into an oscillating dipole. Such oscillating dipole emits light of three different frequencies through Rayleigh scattering, Stokes Raman scattering, and Anti-Stokes Raman scattering.
- ☆ The energy level diagram indicating the different processes and states in the Raman scattering process is shown in **Figure 2**.

3.1. Rayleigh Scattering

- ☆ A molecule with no Raman-active modes can absorb a photon with the energy $h\nu_0$. The excited molecule returns to the ground vibrational state and emits light with the same energy, $h\nu_0$, as excitation light.

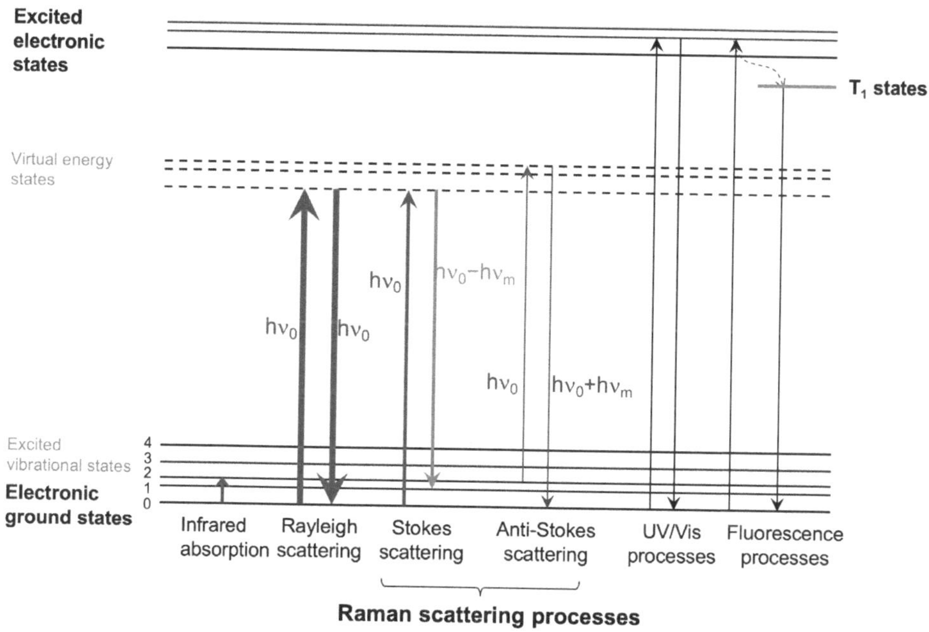

Figure 2: Energy Level Diagram Showing the different Processes and States in the Raman Scattering Process.

The Stokes and anti-Stokes lines are equally spaced from the Rayleigh line. This happens because in both cases, one vibrational quantum of energy is gained or lost. Also, the anti-Stokes line is less intense than the Stokes line, as only molecules that are vibrationally excited before irradiation can produce an anti-Stokes line. Thus, only the more intense Stokes line is usually measured in Raman spectroscopy. Raman scattering is a weak process, and the number of Raman scattered photons is very less. The UV/Vis and fluorescence processes are also shown for comparison. The numbers represent different vibrational levels (vib) within each electronic state.

☆ **Rayleigh scattering:** no change in photon energy after interaction

→ • — •	• — • →
Photon in	**Photon out**
No vibration	***No vibration***

3.2. Stokes Raman Scattering

☆ A photon with energy $h\nu_0$ (and frequency ν_0) can be absorbed by a Raman-active molecule in the ground vibrational state. Part of the photon's energy is transferred to the Raman-active mode with energy $h\nu_m$, and the resulting energy of scattered light is thus reduced to $h\nu_0$-$h\nu_m$. The Raman frequency associated with this energy is called Stokes frequency, or just "Stokes".

☆ **Stokes scattering:** energy lost by photon after interaction

→ • — •	((• — •)) →
Photon in	**Photon out**
No vibration	***Vibration***

3.3. Anti-Stokes Raman Scattering

☆ A photon with energy $h\nu_0$ (and frequency ν_0) can be absorbed by a Raman-active molecule present in the excited vibrational state. Excessive energy of excited Raman-active mode is released, the molecule returns to the ground vibrational state, and the resulting energy of scattered light increases to $h\nu_0+h\nu_m$. The Raman frequency associated with this energy is called Anti-Stokes frequency, or just "Anti-Stokes".

☆ **Anti-Stokes scattering:** energy gained by photon after interaction

→((• — •))	• — • →
Photon in	**Photon out**
Vibration	***No vibration***

☆ Thus, if the incident photon has energy E and vibration energy ν, then in terms of energy, the out photon has energy:

$E - \nu$ = Stokes scattering

$E + \nu$ = Anti-Stokes scattering

E = Rayleigh tokes scattering

☆ About 99.999 % of all incident photons in spontaneous Raman undergo elastic Rayleigh scattering. Such signals are useless for experimental purposes. Only about 0.001 % of the incident light produces an inelastic Raman signal with frequencies $\nu_0 \pm \nu_m$. The spontaneous Raman scattering is very weak that should be distinguished from the predominant Rayleigh scattering. Instruments such as tunable filters, notch filters, laser stop apertures, and double and triple spectrometers have been developed to reduce Rayleigh scattering and obtain high-quality Raman spectra.

☆ Thus, Raman scattering is associated with two processes: Stokes Raman scattering and Anti-Stokes Raman scattering.

☆ The Stokes Raman scattering generates photons with frequencies lower than the excitation source and, thus, are detected at a higher wavelength. While the Anti-Stokes Raman scattering generates photons with frequencies higher than the excitation source and thus are detected at a lower wavelength.

☆ However, in both cases, the Raman shift is dependent on the vibrational structure of the sample.

- The Stokes scattering occurs at lower energy (longer wavelength) than the Rayleigh scattering, and anti-Stokes scattering has higher energy.
- The increase or decrease of energy is related to the vibrational energy levels in the molecule's ground state. Thus, the observed Raman shift of the Stokes and anti-Stokes features provide a direct measure of the vibrational energies of the molecule.
- The energy of the scattered light is less than the incident light for the Stokes lines, and the energy of the scattered light is more than the incident light for the anti-Stokes lines.

4. Principle of Raman Scattering

- Chemical bonds continuously vibrate due to the surrounding thermal energy. Light can interact with these molecular vibrations in predictable ways. Raman spectroscopy uses the scattering of light from molecular vibrations to characterize molecules.
- In Raman spectroscopy, a single wavelength of light is allowed to irradiate a sample. The light is generally a laser in the visible, IR, or UV range (energy range: 200 - 4000 cm^{-1}). Standard Raman wavelengths include 532, 785, and 1064 nm. The sample then scatters the laser light, which is detected by a detector.
- In Raman spectroscopy, monochromatic laser light interacts with molecular vibrations resulting in the increase or decrease in the energy of the laser photons. Such shifting of the energy up or down is called the Raman shift.
- Thus, Raman spectroscopy measures the shift in wavelength of the inelastically scattered light. Raman-shifted photons can be of either higher or lower energy, depending upon the vibrational state of the molecule under study.
- The shift in energy provides valuable information about the vibrational modes in the sample and can help identify the bonds present in the molecule.
- Each compound has a unique Raman spectrum; thus, Raman spectroscopy is an excellent technique for identifying compounds.
- A Raman spectrum consists of several features, each associated with a specific vibrational mode.
- Each chemical bond (*e.g.*, C-H, C-C, C=C) has unique vibrational modes, leading to specific interactions with light, thus producing different Raman shifts (**Table 1**). The Raman shifts can be used to identify molecules and chemical bonds.

Table 1: Raman Shifts (cm^{-1}) for Common Functional Groups

Functional Groups	*Regions*
C-C	~ 600-1300 cm^{-1}
C=C	~ 1600 cm^{-1}
C≡C	~ 2100-2300 cm^{-1}
C-H	~ 2700-3100 cm^{-1}

- The Raman shift depends on the vibrational structure of the molecule **(Figure 2)**. Thus, it is independent of the excitation wavelength, *i.e.*, the chemical fingerprint remains the same.

Note

- Many substances, especially coloured ones, may absorb laser light and emit strong fluorescence, which can contaminate the Raman spectrum. This is one of the major problems in Raman spectroscopy.
- The fluorescence process involves excitation to a real state, S1 (in contrast to the virtual states in Raman spectroscopy). Thus, the fluorescence spectrum is highly dependent on the excitation wavelength.
- To minimize background fluorescence while performing Raman measurements, longer wavelength laser sources are preferred as they cannot provide the sufficient energy required to excite the electron to the S1 state.

- **Raman Efficiency:** The efficiency of Raman scattering is proportional to λ^{-4} (λ is the wavelength of the light). Thus, for an identical sample, the scattering efficiency for a sample excited by a 532 nm laser light will be ~5 times more efficient than a 785 nm laser light and 16 times more efficient than a 1064 nm laser light. This is important while considering the timescales of the experiment, *i.e.*, to achieve the same signal-to-noise ratio, a scan time of 16 times longer is required when using a 1064 nm laser compared to a 532 nm laser.
- **Fluorescence effects:** Fluorescence photons generate in a similar process to Raman scattering but via a different mechanism (see fluorescence spectroscopy). Since, in Raman scattering, the energetic shift relative to the excitation source is measured, the process produces the same spectrum irrespective of the excitation wavelength. In contrast, fluorescence occurs at a fixed wavelength, and thus, it shifts with different excitation wavelengths. Importantly, fluorescence processes are very broad and strong; hence the fluorescence of a sample can suppress the signature Raman features. In addition, darker samples have strong fluorescence signals, so a laser source must be carefully chosen to

minimize fluorescence contributions to the Raman spectrum. A longer wavelength laser light (such as 1064 nm) results in a low fluorescence signal. However, when the wavelength of laser light shifts to the IR region, laser absorption and subsequent sample heating becomes a concern.

- **Sample heating:** Sample heating occurs due to laser absorption, more readily with longer wavelength laser light. The heating can lead to alterations in experimental conditions (*i.e.*, the vibrational spectrum of a sample at different temperatures can be drastically different) and even sample damage (liquid samples may boil and dark-coloured samples can burn or ignite). This can be resolved by lowering the laser, but this will increase the acquisition time for acceptable signal-to-noise.

5. Construction of a Raman Spectrometer

- In a Raman measurement, the sample is illuminated using monochromatic light (in the visible, IR, or UV range). The sample then scatters the laser light, which is collected with a lens and measured by a detector.
- The instrument has the following major components: (1) light source (laser), (2) sample illumination system and light collection optics (lens and mirrors), (3) wavelength selector (filter or spectrophotometer), and (4) detector (photodiode array, CCD, or PMT).
- The double or triple monochromators separate the weak Raman signal from the intense, nearby Rayleigh scattering.

6. Improving Raman Signals

- The Raman signal usually is very weak. Therefore, several methods of sample preparation, illumination, or detection have been developed to enhance the Raman signal intensity, such as:

6.1. Stimulated Raman Scattering

- Instead of using continuous-wave lasers (with electric field ~10^4 V.cm^{-1}), pulsed lasers (with electric field ~10^9 V.cm^{-1}) transform a much larger fraction of incident light into useful Raman scattering and substantially improve the signal-to-noise ratio.
- Stimulated Raman scattering is a type of non-linear Raman spectroscopy.

6.2. Coherent Anti-Stokes Raman Spectroscopy

- Coherent anti-Stokes Raman spectroscopy (CARS) is another non-linear Raman spectroscopy method.
- Instead of using a traditional one laser, light from two strong lasers is used to irradiate a sample. The frequency of the first laser is kept constant, while the frequency of the second is adjusted so that the frequency

difference between the two lasers equals the frequency of some Raman-active mode of interest.

6.3. Resonance Raman Spectroscopy

- If the wavelength of the exciting laser light coincides with the electronic absorption spectrum of a molecule, the intensity of Raman vibrations associated with the absorbing chromophore is enhanced by a factor of 10^2 to 10^4.
- This resonance enhancement or resonance Raman effect can be highly useful in lowering the detection limits and increasing electronic selectivity.
- Thus, the resonance Raman technique can provide structural and electronic insight into species of interest.

6.4. Surface-Enhanced Raman Spectroscopy

- Raman scattering signal is enhanced (typically by 10^3-10^6 times) if the sample is adsorbed on certain colloidal metallic surfaces.
- In SERS, the sample is adsorbed on colloidal metallic surfaces, such as electrochemically *etc.*hed silver electrodes and silver and gold colloids.

6.5. Surface-Enhanced Resonance Raman Spectroscopy

- Surface-enhanced resonance Raman spectroscopy (SERRS) uses both the surface-enhancement effect and resonance Raman effect to enhance the Raman signal intensity as high as 10^{14}.

6.6. Fourier Transform Raman Spectrometer

- The Raman instrument can be performed on the same bench as the FT-IR.
- Here, a 1064 nm YAG:Nd^{3+} laser light is used for excitation so that the excitation energy is lower than the absorption band energies.
- The instrument may also be combined with a microscope, or optical fibre, so that a small surface area (in μ^2) can be scanned, and Raman mapping is easily performed.

7. Applications of Raman Spectroscopy

- Raman spectroscopy is a qualitative tool to identify molecules from their vibrations. It provides a molecular fingerprint representing the set of bonds in the molecule. The vibrational frequencies are characteristic of chemical bonds. The vibrational frequencies depend on the structure and local environment of a molecule, such as morphology, symmetry, crystal phase, solvents, and interactions.
- The quantitative Raman measurements are not sensitive since Raman scattering is weak.

7.1. Study of Protein Structures

- ☆ The precise positions of Raman bands depend on inter- and intra-molecular effects, including peptide bond angles and hydrogen-bonding patterns.
- ☆ The amide I band shows 80 % C=O stretch, near 1650 cm^{-1}. Different secondary structures are characterized by amide I bands slightly different in position and shape.
- ☆ The amide II band shows 60 % N–H bend and 40 % C–N stretch near 1550 cm^{-1}. The amide II band is weak and can only be observed using resonance excitation in Raman spectroscopy.
- ☆ The amide III band shows 40 % C–N stretch and 30 % N–H bend near 1300 cm^{-1}. The amide III bands can be correlated to the amide I band. It provides complementary structural information on the protein structure and, thus, additional details to the amide I band.
- ☆ **Figure 3** and **Table 2** show the amide I, II, and III bands associated with the secondary structure of proteins in Raman spectroscopy.

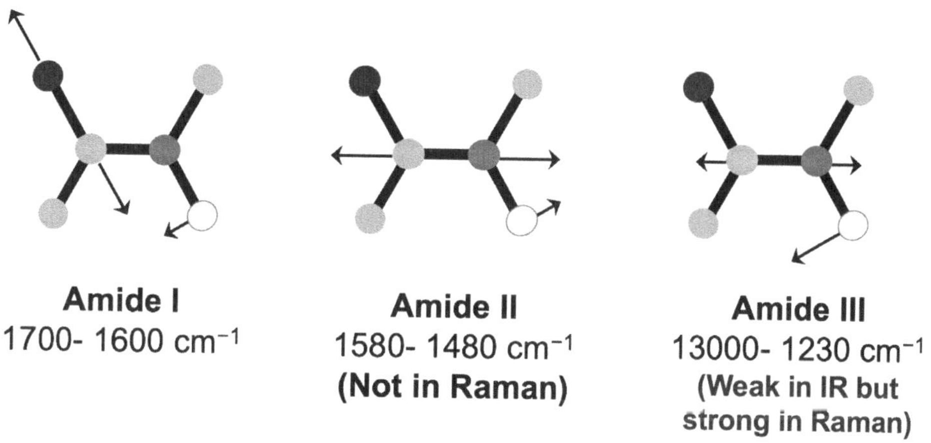

Figure 3: Amide I, II, and III Bands in Raman Spectroscopy.

- ☆ The aromatic amino acids (tryptophan, tyrosine, and phenylalanine) are sensitive to their microenvironment and show Raman vibrational bands.
- ☆ The Raman bands for S-S groups are located in the spectral range of 500-550 cm^{-1}.
- ☆ The Raman spectrum of a protein is influenced by pH, temperature, salts, and solvents. Thus the conformational changes in the protein structure can also be studied using the Raman spectrum, such as protein unfolding, aggregation, *etc.*

Table 2: Amide I, II, and III Band Frequencies and the Secondary Structure of Proteins

Amide Bands	*Wave Number (cm^{-1})*		*Vibration Modes*	*Secondary Structure*
	Average	*Range*		
Amide I	1672	1686–1662	H-bonded C=O stretch	β-turn
Amide I	1654	1660–1648	H-bonded C=O stretch	α-helix
Amide I	1645	1651–1640	H-bonded C=O stretch	Random coil/disordered
Amide I	1633	1640–1626	H-bonded C=O stretch	Parallel β-sheet
Amide I	1626	1640–1612	H-bonded C=O stretch	Anti-parallel β-sheet
Amide III	1280	1300–1270	N-H and C-H bend	α-helix
Amide III	1250	1253–1243	N-H and C-H bend	Random coil/disordered
Amide III	1232	1235–1229	N-H and C-H bend	β-sheet
Skeleton		950–930	N-Cα-C stretch	α-helix

- In addition, the larger the protein, the more complex the Raman spectrum.
- The Raman signal from the peptide groups can be efficiently and selectively enhanced using deep UV resonance Raman spectroscopy (DUVRRS). DUVRRS employs excitation wavelengths in the 190–300 nm range. The peptide bond strongly absorbs at 190 nm, and strong resonances are created using DUVRRS.
- Proteins exhibit low-frequency motions in the form of collective motions of all atoms. These motions are very sensitive to microenvironment variations and can be studied using Raman spectroscopy combined with theoretical calculations such as normal mode analysis.

7.2. Measuring Polarization in Proteins

- In addition to the normal Raman, polarized Raman can provide information about molecular orientation and symmetry of the bond vibrations.
- All molecules are more or less symmetric. Thus, the positioning of these molecules with respect to the laser beam can significantly influence the intensity of Raman bands.
- The excitation of the molecule's different symmetry axes produces different Raman scattering.
- The folded protein structure has local and global symmetries that can be exploited to study molecular polarization using Raman spectroscopy.

7.3. Other Applications

- The influence of chemical reactions associated with proteins, such as folding/unfolding, oxidation-reduction, post-translations modification, *etc.*, can be studied by monitoring the evolution of Raman bands with time. The Raman bands from disulfide bridge stretching, aromatic ring vibrations, and protein side-chain deformation. *etc.*, are followed to understand these phenomena.

- **Figure 4** shows the various information that can be obtained from Raman spectroscopy.

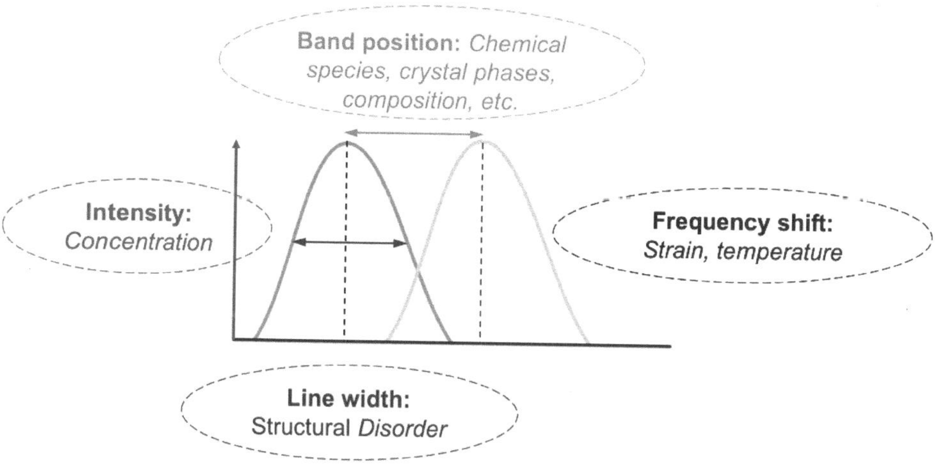

Figure 4: Various Information that can be Obtained using Raman spectroscopy.

- With the development of FT-Raman systems and remote optical fibre sampling, Raman spectroscopy has become a very popular technique.
- Ultra-fast Raman imaging using fluorescence microscopy helps visualize drug delivery, such as the distribution of drug molecules within a cell.
- 3D confocal volume imaging can also be performed to visualize polymeric molecules inside a sample.
- Raman scattering microscopy is also used to image protein aggregates, such as amyloid-beta, inside the cell.
- Raman spectroscopy has also helped in the detection of cancers and the discovery of biomarkers, such as in the case of type2 diabetes.

8. Differences between IR and Raman Spectroscopy

- Both Raman spectroscopy and IR spectroscopy are based on molecular vibrations.
- They are the two most common experimental techniques used to analyze molecular vibrations, motions, and fingerprints.
- Though both are forms of vibrational spectroscopy, IR and Raman spectroscopy differ in certain fundamental aspects.

- Raman spectroscopy is complementary to IR spectroscopy, but sampling is more convenient.
- IR spectroscopy is based on the absorption of light energy corresponding to the vibrational energy of molecules. In contrast, Raman spectroscopy is based on the scattering of incident light at energy shifted by the vibrational energy of the molecule. Vibration modes for the same functional groups are observed at the same wavenumber **(Figure 5)**.

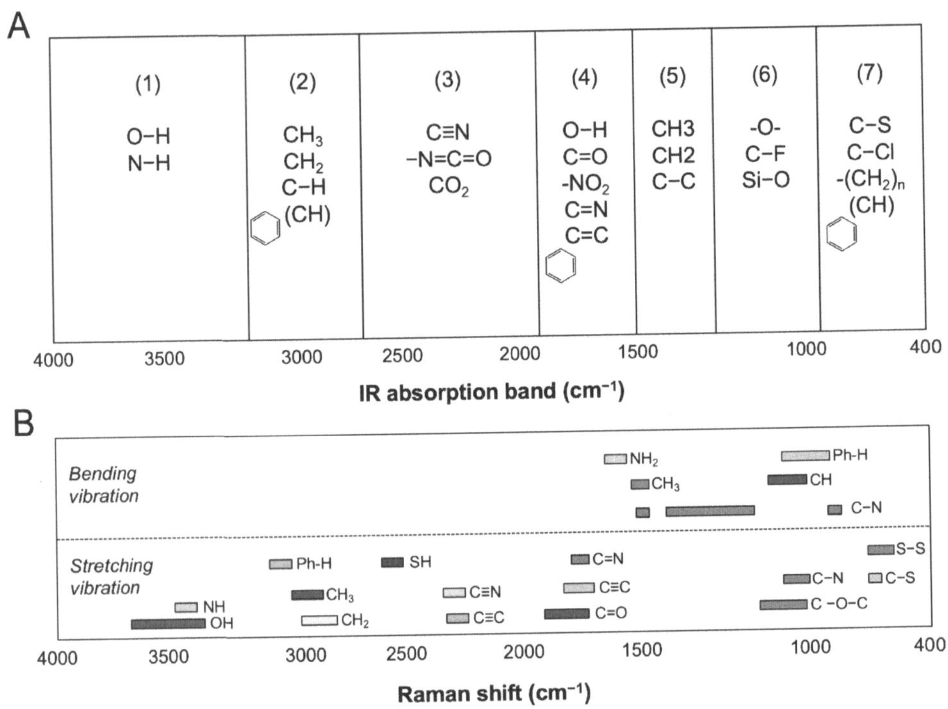

Figure 5: The Difference in the IR and Raman Spectra.
(A) IR Spectrum; (B) Raman Spectrum.

- Some vibrational modes are both IR and Raman active.
- The symmetry of molecules dictates which are active in Raman and IR.
- FT-IR spectroscopy measures the absolute frequency at which a sample absorbs radiation, while Raman Spectroscopy measures the relative frequencies at which a sample absorbs radiation
- FT-IR spectroscopy is sensitive to hetero-nuclear functional group vibrations, polar bonds, *etc.*, while Raman Spectroscopy is sensitive to homo-nuclear and non-polar bonds. For *e.g.*, Raman can distinguish between C-C, C=C, and C≡C bonds.

- ☆ In contrast to FT-IR, Raman Spectroscopy is very amenable to microscopy experiments.
- ☆ The sampling process is more convenient in Raman spectroscopy since glass containers are used.

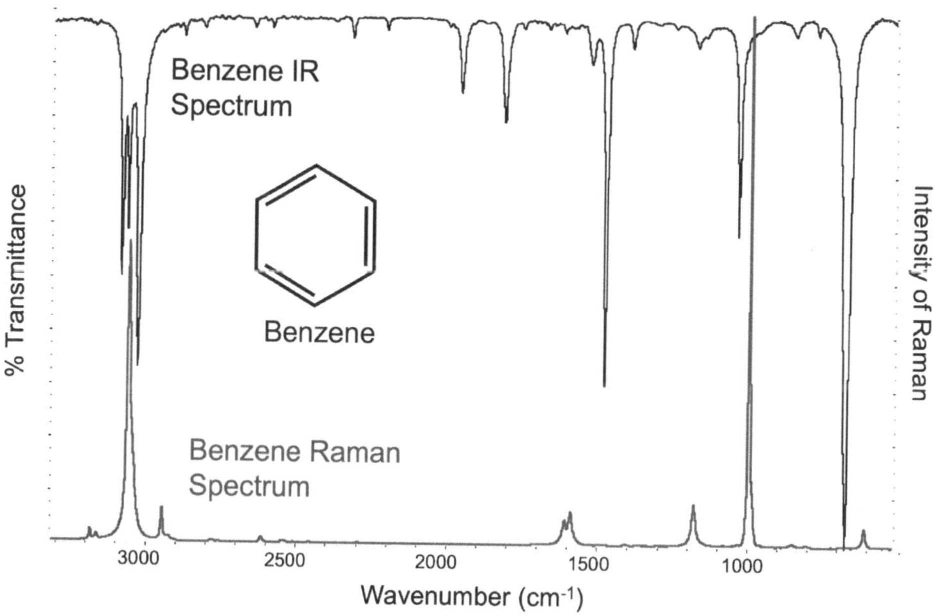

Figure 6: A Comparison of IR and Raman Spectra for Benzene.

- ☆ A comparison of IR transmission and Raman spectra for benzene is shown in **Figure 6**. The intensity of the two spectra exhibits mirror symmetry; thus, the IR and Raman spectra are often considered complementary. But they are different in the type of physical phenomenon they can measure. In IR measurements, the spectral intensity depends on the size of the dipole moment for vibration modes for bonds such as C=O and O-H. On the other hand, in Raman spectroscopy, the intensity depends on the degree of polarizability (electron volume) for vibration modes for bonds such as S-S, C-C, and C≡N.

Chapter 7

Nuclear Magnetic Resonance Spectroscopy

1. Introduction

- ☆ Nuclear magnetic resonance (NMR) spectroscopy is a technique that uses the change in nuclear spin energy under an external magnetic field. Thus, NMR exploits the magnetic properties of specific atomic nuclei to determine the structural properties of any molecule.
- ☆ The principle underlying NMR is that many nuclei have a spin, and all nuclei are electrically charged. If an external magnetic field is applied, an energy transfer is possible between the ground to a higher energy level (generally a single energy gap). The energy transfer occurs at a wavelength corresponding to radio wave frequencies, and when the spin returns to its ground energy level, energy is emitted at the same frequency. The signal that matches this transfer is measured and processed to produce an NMR spectrum for the particular nucleus.
- ☆ In NMR, nuclei (instead of outer electrons) are involved in the absorption process.
- ☆ NMR provides detailed information on the composition, structure, and dynamics of small organic molecules to large molecular weight polymers and proteins.
- ☆ NMR is widely used as the method of choice to determine the chemical structure of organic compounds.

Note

- ☆ What are **N**, **M**, and **R**?

  N = Properties of the **Nucleus,**

  M = The nucleus in a **Magnetic Field,**

  R = When the nucleus meets the right magnetic and radio wave frequency, Nuclear Magnetic **Resonance** occurs.

- ☆ Two common types of NMR spectroscopy that are used to characterise organic structures are:
  1. **^{1}H NMR –** It is used to determine the type and number of H atoms in a molecule.
  2. **^{13}C NMR –** It is used to determine the type and number of C atoms in a molecule.
- ☆ The energy source in NMR is radio waves, which have long wavelengths (~10^7 nm), and thus low energy and frequency.
- ☆ When low-energy radio waves interact with a molecule, they can change the nuclear spins of some elements, like ^{1}H and ^{13}C.
- ☆ In NMR spectroscopy, both liquid and solid samples can be used to study.
- ☆ For liquid samples, conventional solution-state NMR spectroscopy is used, whereas solid-state NMR spectroscopy is used for solid samples.
- ☆ In the liquid phase, different solutions of nucleic acids, proteins, carbohydrates, *etc.*, can be used. In contrast, in the solid phase, samples like crystals, microcrystalline powders, gels, anisotropic solutions, proteins, protein fibrils, various polymers, *etc.*, can be used.

2. Principle of NMR Spectroscopy

- ☆ Nuclei with an odd atomic number or odd mass have nuclear spin. These include ^{1}H and ^{13}C (but not ^{12}C).
- ☆ Since a nucleus is a charged particle in motion (spinning protein), it will have an associated magnetic field.
- ☆ ^{1}H and ^{13}C have a nuclear spin of ½, so they behave like tiny bar magnets.
- ☆ In the absence of a magnetic field, these nuclei are randomly oriented, but when a magnetic field is applied, they line up parallel to the applied field. Their direction can be either spin aligned or spin opposed **(Figure 1)**.
- ☆ NMR spectroscopy is based on the fact that when a population of magnetic nuclei is placed in an external magnetic field, the nuclei align themselves in a predictable and finite number of orientations.

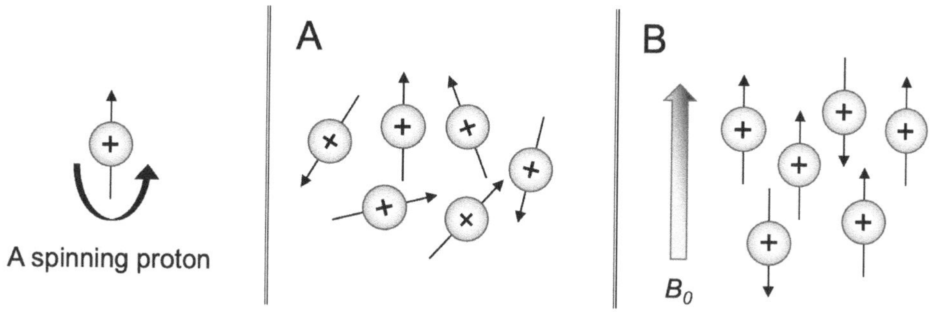

Figure 1: Orientation of Atomic Nuclei in the (A) Absence and (B) Presence of an External Magnetic Field.

A spinning proton creates a magnetic field. (A) In the absence of an external magnetic field (B_0), the nuclei are randomly oriented. (B) In the presence of an external magnetic field, the nuclei are oriented with or against the magnetic field.

2.1 Nuclear Spin

- ☆ Nuclear spin is the total nuclear spin quantum number. It is characterised by a spin quantum number *I*, which may be integral, half-integral, or 0.
- ☆ An electron has a spin quantum number *I* of ½ with allowed values of +½ and -½. This spinning charge creates an associated magnetic field. As a result, an electron behaves like a tiny bar magnet and has a magnetic moment.
- ☆ Any atomic nucleus with an odd mass, an odd atomic number, or both has a property of nuclear spin and a resulting nuclear magnetic moment.
- ☆ The allowed nuclear spin states are determined by the spin quantum number *I* of the nucleus.
- ☆ The number of spin states in a nucleus is $2I + 1$.
- ☆ The most abundant isotopes of carbon (^{12}C) and oxygen (^{16}O) do not have spin.

Table 1: Spin Quantum Numbers of some Atom Nuclei

Element	^{1}H	^{2}H	^{12}C	^{13}C	^{14}N	^{16}O	^{31}P	^{32}S
Nuclear spin quantum number (*I*)	½	1	0	½	1	0	½	0
No. of spin states	2	3	0	2	3	0	2	1

- ☆ All nuclei with a non-zero spin have a magnetic moment, μ, given by

$$\mu = \gamma I$$

where, γ is the gyromagnetic ratio, a proportionality constant between the magnetic dipole moment and the angular momentum specific to each nucleus.

Table 2: The Gyromagnetic Ratios for several Common Nuclei

Nuclei	*Spin*	*Gyromagnetic Ratio (MHz/T)*	*Natural Abundance (%)*
^{1}H	1/2	42.576	99.9985
^{13}C	1/2	10.705	1.07
^{15}N	1/2	-4.361	0.368
^{17}O	5/2	5.772	0.038
^{31}P	1/2	17.235	100

2.2. Splitting of Energy Levels in a Magnetic Field

- ✰ When a charged particle (such as a proton in a nucleus) spins on its axis, it creates a magnetic field (B_0). Thus, the nucleus can be considered to be a tiny bar magnet.
- ✰ Usually, these tiny bar magnets are randomly oriented in space. However, in the presence of an external magnetic field, B_0, the two energy states for a proton align themselves: a lower energy state with the nucleus aligned in the same direction as B_0 and a higher energy state in which the nucleus is aligned against B_0 **(Figure 2)**.

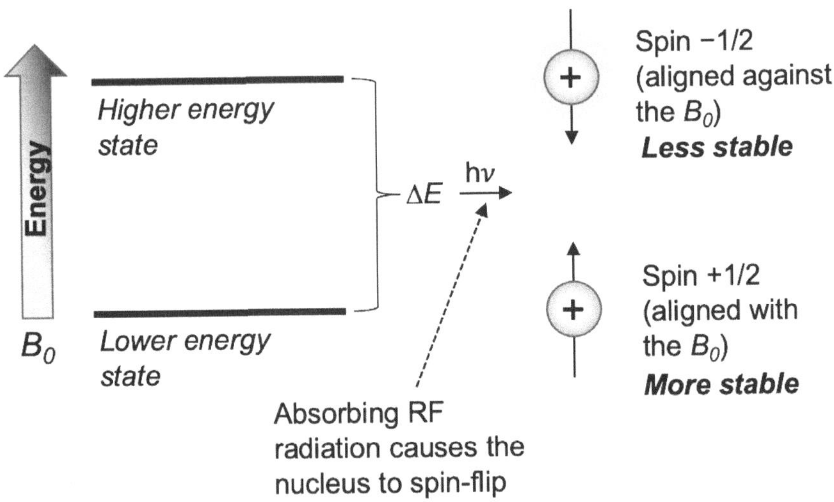

Figure 2: The Orientation of Atomic Nuclei in the Higher and Lower Energy States.

In the presence of an external magnetic field B_0, the two energy states for a proton align themselves: a lower energy state with the nucleus aligned in the same direction as B_0 and a higher energy state in which the nucleus is aligned against B_0.

- When an external energy source that matches the energy difference between these two states is applied, energy is absorbed, causing the nucleus to spin-flip from one orientation to another. Thus, the nucleus is in resonance when it absorbs light of radio wave frequency, and the spin flips to a higher energy state **(Figure 3)**.
- More nuclei are oriented in the same direction as B_0 because it is a lower energy state.
- The energy difference between these two nuclear spin states is very low (<0.1 cal). It corresponds to the low radio wave frequency region of the electromagnetic spectrum.
- For ^1H and ^{13}C, only two orientations are allowed.
- Two variables characterise NMR: (i) an applied magnetic field B_0, the strength of which is measured in tesla (T), and (ii) the frequency ν of radiation used for resonance, measured in hertz (Hz), or megahertz (MHz).
- The frequency needed for resonance (ν) and the applied magnetic field strength (B_0) are proportionally related:

$$\nu \propto B_0$$

- The stronger the magnetic field, the larger the energy difference between two nuclear spin states and the higher the frequency needed for the resonance.

2.3. Spin-spin Interaction

- If a nucleus with a spin ½ is placed in a magnetic field, the nucleus exists in the lower energy level (*i.e.*, its magnetic moment does not oppose the applied magnetic field) while the nucleus is spinning on its axis.
- In the presence of a magnetic field, this axis of rotation will *precess* around the magnetic field and traces out a cone-shaped surface in the same way as a spinning top trace out a cone-shaped surface (Precession is the change in the orientation of the rotational axis of a rotating body) **(Figure 3)**.
- The rate of precession is expressed as a frequency in hertz.
- The nuclear magnetic moment will couple to the applied external magnetic field B_0, which produces a torque on the nucleus and causes the precession around the magnetic field. This is analogous to the macroscopic tops in that the gravitational force couples with the mass of the top. The frequency of precession is known as the Larmor frequency, ν_0.

$$\nu_0 = \gamma B_0$$

where, γ is the gyromagnetic ratio.

- ☆ If the precessing nucleus is irradiated with electromagnetic radiation of the same frequency as the rate of precession, then the two frequencies couple, energy is absorbed, and the nuclear spin is flipped from spin state +½ (with the applied field) to -½ (against the applied field) **(Figure 3)**.

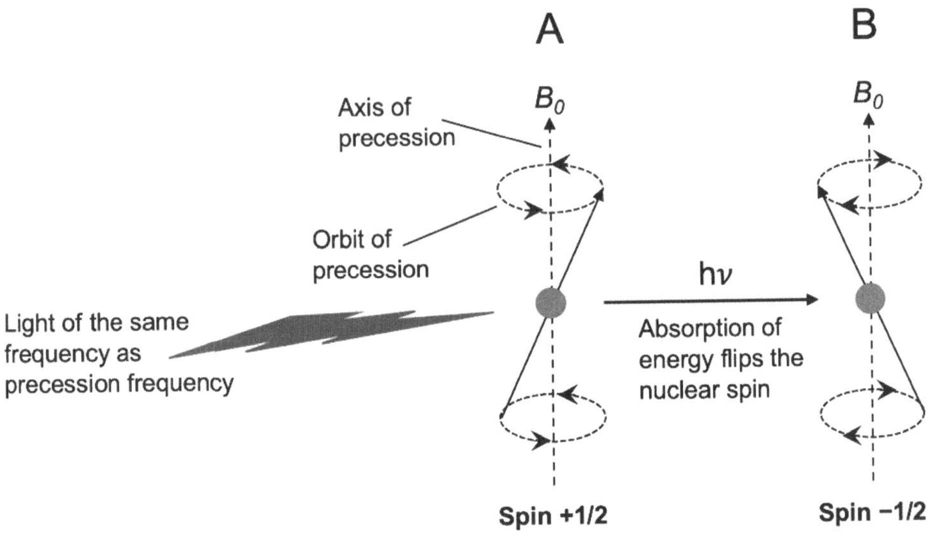

Figure 3: Origin of the Nuclear Magnetic Resonance.

(A) Precession of a spinning nucleus in an applied magnetic field. (B) Absorption of light occurs when the frequency of radiation is equal to the frequency of precession.

- ☆ The NMR experiment measures the resonant frequency that causes a spin flip when ν matches ν_0.

2.4. Relaxation Processes

- ☆ Resonance is the absorption of electromagnetic radiation by a precessing nucleus and the resulting flip of its nuclear spin from a lower to a higher energy state.
- ☆ The NMR spectrometer detects the coupling of precession frequency and electromagnetic radiation and records it as a signal.
- ☆ The process is quantised so that only electromagnetic radiation of the correct frequency causes nuclear spin to flip.
- ☆ Electromagnetic radiation with a very high or very low frequency is not absorbed.

2.5. Shielding and Deshielding

- ☆ Not much information could be obtained if all protons absorb the same amount of energy in a given magnetic field.

☆ However, protons are surrounded by electrons that shield them from the external field **(Figure 4)**.

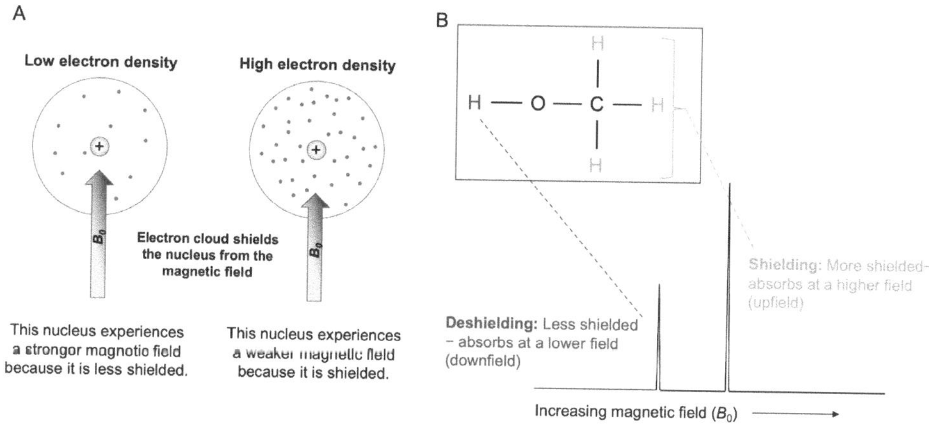

Figure 4: Nuclear Shielding Effect.

(A) A nucleus is called shielded if its chemical shift has been decreased due to the addition of electron density or magnetic field. A nucleus is called deshielded if its chemical shift has been increased due to the removal of electron density or magnetic field. (B) NMR spectrum showing applied field strength increasing from left to right. The left part is downfield, and the right is upfield. Nuclei that absorb on the upfield side are strongly shielded, whereas nuclei that absorb on the downfield side are weakly shielded.

☆ Circulating electrons create an induced magnetic field that opposes the external magnetic field. Thus, the magnetic field strength must be increased for a shielded proton to flip at the same frequency.

☆ Depending on their chemical environment, protons are shielded by different amounts.

2.6. Chemical Shift

☆ The relative energy of resonance of a particular nucleus resulting from its local environment is called the chemical shift. Thus, it is a function of the nucleus and its environment.

☆ The chemical shift indicates the strength of the magnetic field that the nucleus experiences. It is defined as nuclear shielding/applied magnetic field.

☆ The chemical shift is an extremely important value as it provides information about the local structure surrounding the nucleus of interest.

☆ The chemical shift is measured relative to a reference compound. For ^1H NMR, the reference is usually tetramethylsilane [Si $(CH_3)_4$].

☆ Tetramethylsilane protons are highly shielded (since silicon is less electronegative than carbon), and the signal is taken as zero.

- The electrons surrounding a nucleus affect the effective magnetic field sensed by the nucleus. More shielded nuclei sense a smaller effective magnetic field, while the less shielded nuclei sense a larger effective magnetic field. As a result, more shielded nuclei come into resonance at a lower frequency, while the less shielded nuclei go into resonance at a higher frequency **(Figure 4)**.
- The energy difference between the two spin states is much smaller in energy for more shielded nuclei and resonates at a lower frequency. Less shielded nuclei have a much higher energy difference between the two spin states, which resonate at a much higher frequency **(Figure 5)**.

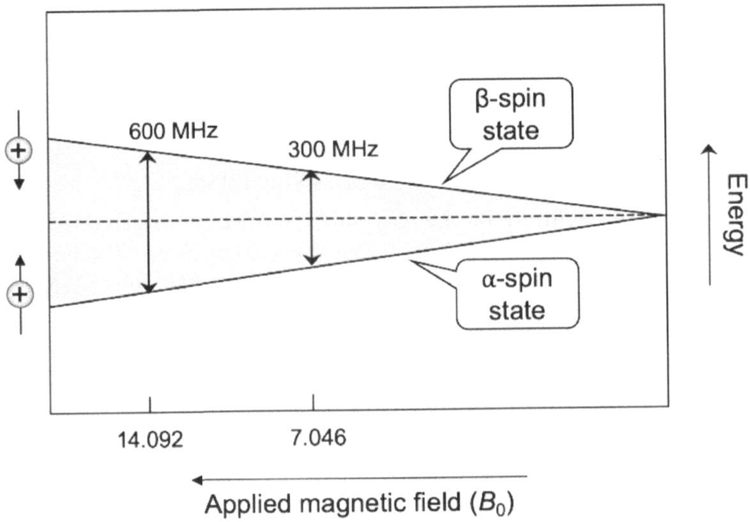

Figure 5: Energy difference between the Two Spin States.

Note

- More electronegative atoms deshield more and give larger shift values.
- The effect decreases with distance.
- The presence of additional electronegative atoms causes an increase in chemical shifts.

2.6.1. Measurement of Chemical Shifts

- The numeric value of chemical shift (δ) is the difference between the strength of the magnetic field at which the observed nucleus resonates and the field strength for the resonance of a reference. The difference is minimal but can be accurately measured.
- It is taken as a ratio to the total field multiplied by 10^6, so the shift is in parts per million (ppm).

$$\delta = 10^6 \times \frac{\nu - \nu_{TMS}}{\nu_{TMS}}$$

where, ν (in Hz) is the frequency of the interested line, and ν_{TMS} is the frequency of the line from TMS.

☆ Absorptions usually occur downfield of tetramethylsilane to the left on the chart.

2.6.2. NMR Signals

☆ The **number** of signals shows how many different kinds of protons/ carbons are present.

☆ The **location** of the signals shows how shielded or deshielded the proton is (*i.e.*, chemical shift indicates the type of functional group).

☆ The **intensity** of the signal indicates the number of protons/carbons.

☆ The signal **splitting** pattern shows the number of protons attached to the adjacent carbon atoms.

2.7. Construction of an NMR Spectrometer

☆ The main components of an NMR spectrometer are (1) a powerful magnet, (2) a radio-frequency generator, (3) a radio-frequency detector, (4) a sample tube, and (5) a computer.

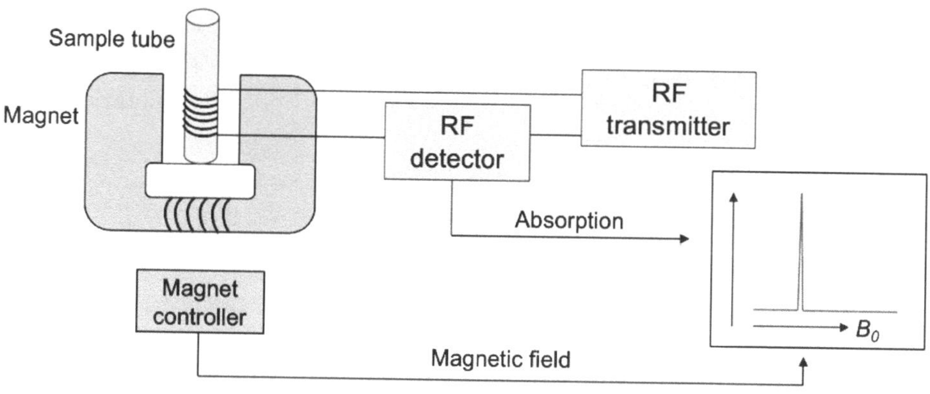

Figure 6: Schematic Representation of an NMR Spectrometer.

3. Working of an NMR Spectroscopy Experiment

☆ The sample is dissolved in a solvent, most commonly CCl_4, $CDCl_3$, or D_2O, which does not have any ^1H and thus does not interfere with ^1H spectra. The sample is placed in a glass sample tube.

☆ A radio-frequency generator then irradiates the sample with a short pulse of radiation, causing resonance.

- ☆ When the nuclei fall back to their lower energy state (from +½ to -½ spin state), the detector measures the energy released, and a spectrum is recorded.
- ☆ Protons in different environments absorb at slightly different frequencies, making them distinguishable by NMR. The frequency at which a particular proton absorbs is determined by its electronic environment.
- ☆ The size of the magnetic field generated by the electrons around a proton determines where it absorbs. The absorption frequencies are plotted as peaks relative to the tetramethylsilane reference on a ppm scale.
- ☆ The received NMR response is very weak in signal and requires a sensitive radio receiver to pick up. A Fourier transform is done to extract the frequency-domain spectrum from the raw time-domain spectrum.
- ☆ Good ^{1}H NMR spectra can be acquired with ~16 repeats, which takes only a few minutes. However, for heavier elements than ^{1}H, acquiring quantitative heavy-element spectra can be time-consuming, taking a few minutes to hours.
- ☆ An average of all the acquired spectrums is generated and displayed as a graph.

Note

- ☆ New Fourier transform NMR (FT-NMR) spectrometers are very sensitive and fast equipment; a spectrum can be recorded in a few seconds.
- ☆ Modern NMR spectrometers use a constant magnetic field strength B_0, and then a narrow range of frequencies is applied to achieve the resonance of all protons.
- ☆ Only nuclei that contain odd mass numbers (such as ^{1}H, ^{13}C, ^{19}F, and ^{31}P) or odd atomic numbers (such as ^{2}H and ^{14}N) give rise to NMR signals.

4. ^{13}C NMR Spectroscopy

- ☆ ^{13}C is the only carbon isotope with a nuclear spin. In organic molecules, the natural abundance of ^{13}C is 1.1 % (^{12}C accounts for more than 98 % of the carbons).
- ☆ The ^{13}C nucleus has a much weaker magnetic moment than a ^{1}H. As a result, the NMR signals from ^{13}C nuclei are much weaker than proton signals. This, combined with the low natural abundance of ^{13}C, makes detecting carbon signals more challenging with a much lower signal-to-noise ratio than in ^{1}H NMR. Therefore, more concentrated samples are required to generate a useful spectrum. Additionally, to bring the signal-

to-noise ratio down to acceptable levels, the data from hundreds of scans are averaged.

- All signals are obtained simultaneously using a broad pulse of energy and resonance recorded.

5. ^{1}H NMR Spectroscopy

- Proton or ^{1}H NMR is much more sensitive than ^{13}C, as the active ^{1}H nucleus is nearly 100 % of the natural abundance.
- The number of signals in the NMR spectrum corresponds to the number of non-equivalent proton groups in the molecule.
- The knowledge of how to identify proton equivalence help to identify non-equivalent proton groups and identify the number of signals that will show up in the NMR spectrum for the molecule.
- Equivalent hydrogens have the same signal, while non-equivalent are "different" and, as such, may cause additional splitting (diastereotopic effect).
- Equivalent protons: protons that are identical in every way. They have the same magnetic environment, which gives them the same spin-flip energy.
- Non-equivalent protons: protons that only differ in one way. These protons do not have the same magnetic environment.

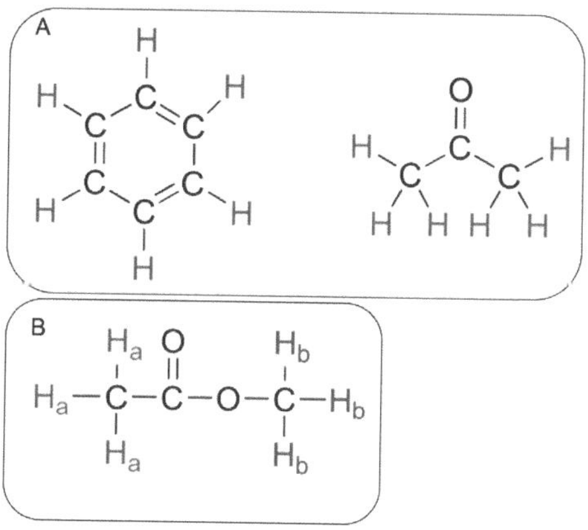

Figure 7: Equivalent and Non-equivalent Protons.

(A) In benzene and acetone, all protons are chemically equivalent and will have the same resonance frequency. (B) Methyl acetate contains two sets of protons. The resonance frequency of the three H_a protons will differ from that of the three H_b protons because the two sets of protons are in non-identical environments (*i.e.,* chemically non-equivalent). It should be noted that the three H_a protons are all in the same electronic environment (chemically equivalent to each other) and will have identical resonance frequencies.

☆ A molecule with 1 set of equivalent hydrogens gives one NMR signal. In contrast, a molecule with two or more sets of equivalent hydrogens gives a different NMR signal for each set.

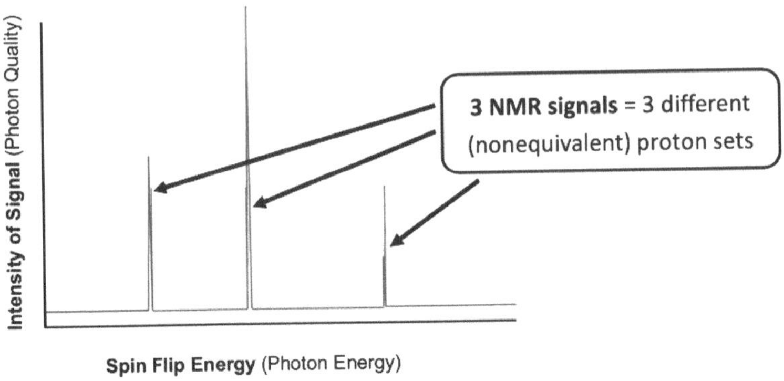

Figure 8: NMR Signals for a Molecule with Three Sets of Protons.

☆ Theoretical equivalence can be predicted by seeing if replacing each H with "X" gives the same or different outcome, as replacing each H with "X" shows a different constitutional isomer. The hydrogens are then in constitutionally heterotopic environments and will have different chemical shifts – they are non-equivalent under all circumstances.

6. Example of a ^{13}C and ^{1}H Spectra

6.1. ^{1}H NMR Spectrum of Ethanol

☆ There are three chemical shift lines in the ^{1}H NMR spectrum of ethanol, indicating three different types of protons/hydrogens (**Figure 9A**).

☆ The relative shielding of the hydrogen atoms on methyl and methylene groups is responsible for the differences in chemical shifts.

☆ Since CH_2 is directly attached to a more electronegative atom Oxygen, so, the signal due to CH_2 is shielded from the electromagnetic field compared to the CH_3 protons. As a result, CH_3 protons appear upfield compared to CH_2.

☆ The splitting of the peaks for CH_2 (quartet) and CH_3 (triplet) is due to the spin-spin splitting, the effect of neighbouring protons (n+1 times, n is the total number of neighbouring protons/hydrogens), which can be obtained with a high-resolution NMR spectrometer **(Figure 9B)**.

☆ Splitting is caused by the spins of protons on adjacent carbon atoms coupled with each other.

☆ For the CH_3 group, n=2, so multiplicity is 3 (triplet); for the CH_2 group, n=3 so multiplicity is 4 (quartet).

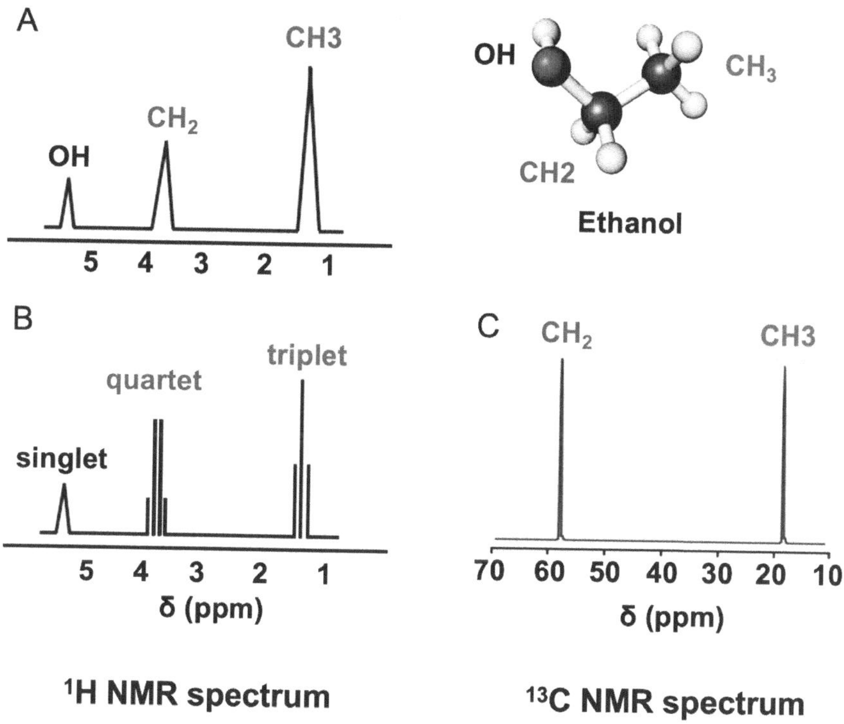

Figure 9: ^{1}H and ^{13}C NMR Spectra of Ethanol.

(A) ^{1}H NMR Spectrum of Ethanol with Characteristic Peaks. (B) spin-spin splitting of peaks with high-resolution spectrum. (C) ^{13}C NMR spectrum of ethanol.

- ☆ Similarly, in the -OH group proton should appear as a triplet. However, due to the fast interchange of the -OH proton with other solvent protons. It usually appears as a broad peak (singlet) at around 5 ppm.
- ☆ From ^{1}H NMR spectra, the number of sets of peaks indicates the number of different hydrogen environments in a compound, and the area under each set of peaks indicates the number of hydrogens.

6.2. Characteristic ^{13}C NMR Spectrum of Ethanol

- ☆ There are two chemical shift lines in the ^{13}C NMR spectrum of ethanol, indicating two different chemical environments of carbon atoms **(Figure 9C)**.
- ☆ The external magnetic field experienced by the carbon nuclei is affected by the electronegativity of the atoms attached to them.
- ☆ The chemical shift of the carbon increases when an atom like oxygen is attached to it. This means that the peak at about 60 ppm (the larger chemical shift) is due to the CH_2 group because it has a more electronegative atom attached.

7. Applications of NMR

- NMR spectroscopy uses the NMR phenomenon to study the physical, chemical, and biological properties of matter.

7.1. Determination of Molecular Structures of Chemical Entities

- NMR spectroscopy is the primary method of structure determination for organic compounds.
- It is used in conjunction with mass spectroscopy and infra-red spectroscopy.
- It has wide applications in natural product chemistry and synthetic organic chemistry.
- The chemical shift provides a clue about the environment of a particular proton or carbon, thus allowing an understanding of the functional groups.
- Spin–spin interactions allow an understanding of how protons are linked with the carbon skeleton.
- The signal intensity also explains how many protons contribute to a particular signal.

7.2. Study of the Solution Structure of Proteins and Peptides

- Determination of 3-D structures of proteins, protein-ligand complexes, DNA, RNA, protein/DNA complexes, polysaccharides, *etc.*
- The structures of proteins up to a mass of about 50 kDa can be determined with biomolecular NMR spectroscopy.
- Preparing proteins for NMR requires recombinant expression and isotopic labelling to enrich the samples with ^{13}C, ^{15}N, or ^{1}H labelling.
- Assignment of chemical shifts and interatomic distances is carried out with the help of software programs.

7.3. Magnetic Resonance Imaging

- The basic principles of NMR can be applied to the imaging of live samples.
- Magnetic resonance imaging (MRI) helps to form pictures of the anatomy and the physiological processes of the body in both health and disease.
- Certain atomic nuclei can absorb and emit radiofrequency energy when placed in an external magnetic field.
- In MRI, ^{1}H atoms are most often used to generate a detectable RF signal that is received by antennas close to the anatomy being examined.
- ^{1}H is naturally abundant in biomolecules, particularly in water and fats.
- Thus, most MRI scans map the location of water and fat in the body, creating a signal that is processed to form an image of the body.

- ☆ Pulses of radio waves excite the spin energy transition, and magnetic field gradients localise the signal in space.
- ☆ By varying the parameters of the pulse sequence, different contrasts may be generated between tissues based on the relaxation properties of the hydrogen atoms therein.

7.4. Study of Dynamic Processes

- ☆ NMR can be used to study reaction kinetics, equilibrium (chemical or structural), protein folding, *etc.*

7.5. Drug Design and Discovery

- ☆ NMR is widely used in drug discovery and development.
- ☆ NMR can detect transient and weak interactions; thus, becoming a powerful tool in drug discovery.
- ☆ NMR is also used to study the structure-activity relationships of enzyme-drug complexes.

Note

- ☆ In the case of large molecular weight proteins, there is an overlap of peaks in the NMR spectrum. In such cases, one can utilise the excellent resolution and sensitivity offered by ^{13}C and ^{15}N isotopes combined with multidimensional NMR experiments. Proteins can be enriched with ^{13}C or/and ^{15}N isotopes.

Chapter 8

X-ray Spectroscopy

1. Introduction

- ☆ X-rays were discovered by William Conrad Röntgen in 1895, for which he was awarded the Nobel Prize in Physics in 1901. He was the first recipient of a Nobel Prize.
- ☆ X-ray spectroscopy deals with the absorption and emission of X- rays by a matter.
- ☆ X-ray spectroscopy is a general term used for several spectroscopic techniques to characterize samples using X-rays.
- ☆ Crystal X-ray spectroscopy or X-ray crystallography uses the technique of X-ray spectroscopy to determine the position and arrangement of atoms in a crystal.
- ☆ X-ray crystallography is a powerful technique for visualizing the atomic and molecular details of molecules from crystals.
- ☆ It is the most commonly used method for determining the three-dimensional (3D) structure of biomolecules like proteins or nucleic acids.

Note

- ☆ X-ray crystallography, nuclear magnetic resonance (NMR) spectroscopy, and cryo-electron microscopy (cryo-EM) are the most commonly used methods for three-dimensional biomolecular structure determination.

- ☆ When an X-ray hits an object, the object's atoms scatter the X-ray beams.
- ☆ Max von Laue discovered that X-rays passing through a crystal interact with it and, as a result, diffract in particular directions depending on the nature of the crystal. For this, they were awarded the Nobel Prize in Physics in 1914.
- ☆ Since crystals have a regular arrangement of atoms, they diffract the X-rays in specific directions. By measuring the directions and the intensity of the scattered beams, the 3D atomic structure of the crystal can be studied.
- ☆ Crystals are ideal subjects for studying the macromolecular structures at the atomic level, as they have three specific characteristics: i) they are solids, ii) they are three-dimensional, and iii) they are built from very regular and highly symmetrical arrangements of atoms.
- ☆ William Henry Bragg and William Lawrence Bragg (1913) discovered that X-rays could be used to accurately determine the positions of atoms within a crystal and unravel its 3D structure. X-rays are diffracted by planes within the crystal, according to Bragg's law. The diffracted X-rays are collected as dark spots on an image plate. This diffracted pattern provides information about the structure of the crystal. Bragg's law formed the basic principle of modern biomolecular structure determination. For this discovery, the father and son team (W. H. Bragg and W. L. Bragg) were awarded the Nobel Prize in Physics in 1915.
- ☆ John Kendrew and Max Perutz used X-ray crystallography and were the first to solve the crystal structure of the muscle protein myoglobin (1959). For this, they were awarded the Nobel Prize in Chemistry in 1962.

2. Properties of X-rays

- ☆ The X-rays are electromagnetic radiations with wavelengths between 0.02 and 100 Å corresponding to energies from 100 eV to 100 keV.
- ☆ As the wavelength of X-rays falls in the range of atomic bond distances, they can be used for determining atomic positions in a molecule and thus help study biological structures. The resulting diffracted rays are collected as dark spots on an image detector. This diffracted pattern provides information about the structure of the crystal.
- ☆ The wavelength of X-rays is smaller than visible light; thus, they have higher energy than visible light.
- ☆ X-rays can penetrate matter more easily than visible light as they have higher energy. Their capacity to penetrate matter depends on matter density, and thus X-rays are a powerful tool used in medicine for mapping the internal structures of the human body. Bones have a higher density than tissues and, thus, are harder for X-rays to penetrate. Fractures in bone have a different density than the bone; therefore, fractures can be seen in X-ray images.

☆ A molecule can ionize when molecules are excited with X-rays (high-energy and short wavelength radiation). When an electron of an atom is excited by a photon's energy, it jumps from a lower energy level to a higher energy level. When the electron returns to the lower energy level, the energy previously gained during excitation is emitted as a photon. This photon has a wavelength/energy characteristic for the element (there can be several characteristic wavelengths per element). Thus, X-rays are emitted during electronic transitions to the inner shell states in atoms of modest atomic numbers. These X-rays have specific energies associated with the atomic number. Thus, each element has a typical X-ray spectrum that can be used for its identification.

3. Steps in X-ray Crystallography

☆ X-ray crystallography is among the most vital techniques in physical, chemical, and biological science because it is one of only a few methods to determine the structure of large biomolecules directly. However, the steps in X-ray crystallography are tedious, time-consuming, and require a skilful hand. The overall steps to determine the structure of a protein using X-ray crystallography are shown in **Figure 1**.

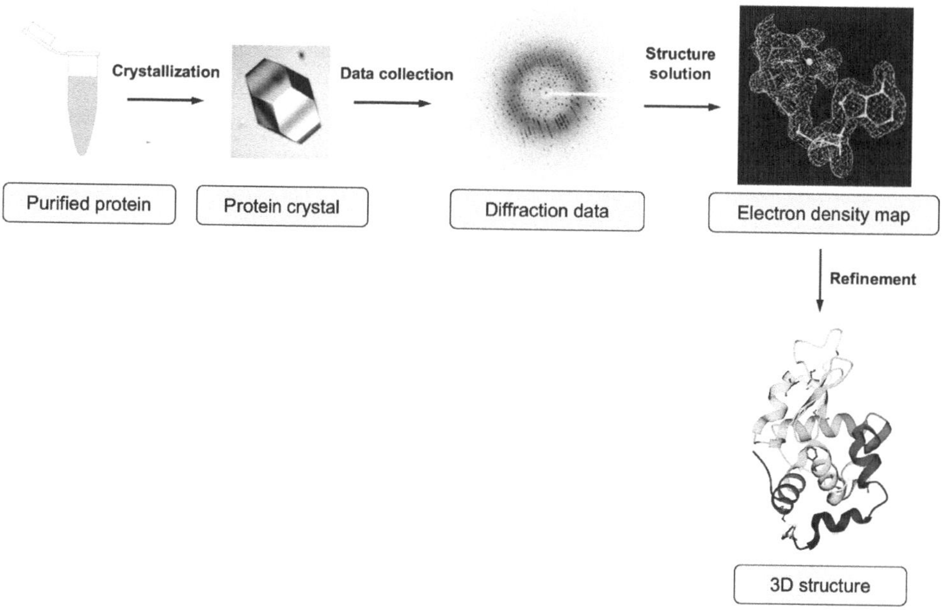

Figure 1: Steps in Protein Structure Determination using X-ray Crystallography.

1. The first step in protein crystallography is the production of pure protein. This is achieved using various chromatographic techniques.

2. The purified protein is used to form crystals. The process of making crystals is known as crystallization. There is no method to predict the crystallization of any molecule. To grow crystals of any compound from the solution, the molecules must be brought to a supersaturated state. When the solution returns from a supersaturated state to equilibrium, the substance is precipitated, hopefully as crystals and not as an amorphous precipitate. A crystal is formed by the periodic assembly of molecules in a 3D array that develops by the separation of solvents from the molecules in the solution. Supersaturation can be achieved by slow evaporation. Also, by varying one of the factors affecting crystallization, one may change the solubility and hence the saturation level of the molecule.
3. The X-ray beam is exposed to the crystal, and the diffraction data is collected in the form of spots.
4. The integrated intensities of the diffraction spots are used to reconstruct the electron density map. The reconstruction is done by a Fourier transformation of the diffraction intensities with phase angle assignment. A high degree of completeness and redundancy in diffraction data is necessary, *i.e.*, all possible reflections are measured multiple times using a detector that can collect diffraction data. The use of high-intensity X-ray sources, such as synchrotron radiation, is a new and better way to collect X-ray diffraction data with greater accuracy.
5. The atoms are placed in the observed electron density, and their positions are refined to get the 3D structure of the molecule.

Note

- ☆ A crystal is a regular, repeating 3D array of atoms or molecules or groups of molecules.
- ☆ Crystals are built from strong covalent or ionic bonds; thus, biological crystals are solid, soft, and brittle.
- ☆ Crystallization is the rate-limiting step in crystallography.
- ☆ X-ray scattering by a single molecule will be extremely weak and noisy, and thus difficult to detect and analyze. On the other hand, in crystals, a large number of molecules are in the same orientation, and the signal gets added up and raised tremendously. The crystal acts as an amplifier. Hence a crystal is required to observe the measurable intensity of the scattered X-rays.

4. Construction of an X-ray Spectrometer

- ☆ An X-ray spectrometer is also called an X-ray diffractometer.
- ☆ Figure 2 shows the main components of an X-ray spectrometer that are (1) X-ray generator, (2) Goniometer, (3) X-ray detector, and (4) Computer.

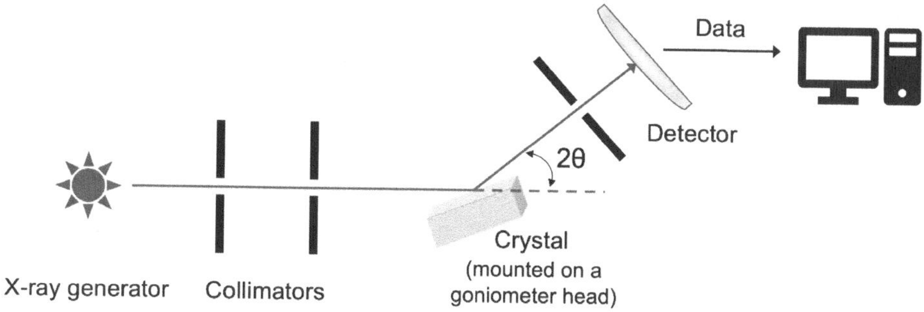

Figure 2: Schematic Representation of an X-ray Spectrometer.
Details are discussed in the main text.

4.1. X-ray Generator

- The X-ray generator is kept in a vacuum tube.
- To generate X-rays, the electrons are released by a hot cathode and accelerated by an electric field (about 30 keV) against a metal target (the anode). The high-velocity electrons collide with a metal target, creating X-rays.
- The commonly used metal targets are made up of either copper or molybdenum.

Note

- A synchrotron is a new and extremely powerful source of X-rays.
- In a synchrotron, the X-rays are produced by high-energy electrons as they circulate the synchrotron orbit in storage rings.
- The storage ring at the European Synchrotron facility is an 844 meters circular tube where the electrons circle close to the speed of light. The tube is maintained at ultra-low pressure. As the electrons travel around the ring, they pass through different magnets and, in the process, produce X-rays.
- A synchrotron machine accelerates electrons to extremely high energy and then makes them change direction periodically. The resulting X-rays are emitted as dozens of thin beams, each directed toward a beamline next to the accelerator.
- The development of synchrotrons has revolutionized crystallography.

4.2. Goniometer

- The most important component of the X-ray diffractometer is the goniometer assembly. The goniometer allows the crystal to be rotated and precisely oriented in any direction while the X-ray beam is focused on the crystal.

- The goniometer assembly must be capable of keeping the crystal centred in the incident X-ray beam while, at the same time, changing its orientation in order to collect thousands of frames of data.
- At the head of the goniometer, the crystal is mounted. The crystal is attached to the glass fibre tip by an adhesive, and the goniometer head is placed at the base of the goniometer assembly.
- The crystal is optically aligned in the centre of the incident X-ray beam using a video camera or microscope.

4.3. X-ray Detector

- The most important component of any modern X-ray diffractometer system is its detector system.
- The first X-ray detector used was a photographic film.
- Currently, several types of detectors are available, such as scintillation counters, area detectors, charge-coupled devices, image plates, *etc.*
- Area detectors are the most commonly used detectors in data collection techniques.
- Most new generation X-ray diffractometers use detector systems based on charge-coupled device (CCD) technology. These detectors include a CCD microprocessor chip to record frame intensities.
- Image plates are also used as detectors. Image plates are exposed to X-rays, and the X-ray photon causes a chemical change in the plate coating that releases a fluorescence, which is detected by a photomultiplier when scanned with the light of the correct wavelength. Image plates are read out on a scanner. Image plates have a wider range of sensitivity with respect to X-ray wavelength, which gives them higher counting efficiency at higher energies.

4.4. Computer

- A computer is used for data collection, calculations, and structure determination.
- Modern X-ray diffractometers have sophisticated software packages (CPP4 package https://www.ccp4.ac.uk/) to solve the structure using computer programs.

5. Working of an X-ray Diffractometry Experiment

- The most common method of X-ray crystallography is single-crystal X-ray diffraction, in which the atoms present in the crystal cause the incident X-rays to produce scattered rays. When the scattered beams fall on the detector, they create a speckle diffraction pattern.

- ☆ When a sample is illuminated by an intense X-ray beam, known as the incident beam, some energy is scattered, but some are absorbed by the sample depending on its structure.
- ☆ As the crystal is gradually rotated, the angle and intensity of these diffracted beams can be measured from each set of planes, and a 3D image of the electron density can be generated.
- ☆ Based on this electron density, the average position of atoms in the crystal, chemical bonds, and various information can be determined.
- ☆ The intensity of the energy is directly proportional to the abundance of the element in the sample. The exact value for each element can be determined after comparing it with the available standards.

Note

- ☆ If a sample has many elements, the wavelength-dispersive spectrometer can be used to segregate a complex resultant X-ray spectrum into specific wavelengths for each element present.

6. Applications of X-ray Spectroscopy

6.1. To Study Macromolecular Structures

- ☆ X-ray spectroscopy is an important method for determining the atomic resolution structure of small compounds. When other spectroscopic methods fail to reveal a compound's identity, X-ray spectroscopy is the method of choice for structural determination, as it can determine accurate 3D structural information of the compounds, such as bond lengths and bond angles.
- ☆ X-ray spectroscopy is a method of choice for determining biomolecular structures, particularly proteins and nucleic acids. As of early 2023, around 200,000 structures of proteins, nucleic acids, oligosaccharides, and their complexes have been determined using X-ray spectroscopy. However, the technique is limited in that it requires biomolecules in the form of a crystal. The crystallization process is tedious, time-consuming, and often unpredictable. In several cases, the time spent searching for crystallization conditions takes years.

6.2. Drug Design and Discovery

- ☆ X-ray crystallography is the most common technique used for structure-based drug design. The protein structure with high-resolution details of surface amino acids, stereochemistry, and active site geometry can be determined. Such data can be used to design small ligand compounds with high affinity and selectivity to inhibit the activity of the protein. Further, these compounds can be used as lead candidates for *in vitro* inhibition assays and clinical trials.

6.3. To Study Disease Mechanisms

- X-ray crystallography has been widely used to study the cause and mechanism of various diseases, such as sickle cell anaemia, thalassemia, amyloidosis, *etc.* If a protein is formed due to a faulty gene, mutation, or other errors, it will show structure with the native protein. Such analysis may provide insights into the mechanism of the disease and the treatment possibilities.

6.4. Other Applications

- X-ray spectroscopy is also used for certain analytical purposes, such as in the fields of metallurgy, geology, mineralogy, ceramics and glass manufacturing, *etc.* It is also used in several environmental studies (*e.g.*, analyses of particulate matter on air filters) and in the petroleum industry (*e.g.*, sulfur content of crude oils and petroleum products). The X-ray analytical methods are non-destructive.

7. Comparison of X-ray Crystallography and NMR Spectroscopy

- The 3D structure of proteins provides insights into their biological activities and disease mechanisms. Understanding the structure also allows the design of new drugs.
- There are three primary techniques in structural biology: X-ray crystallography, NMR spectroscopy, and cryo-EM. However, each technique has its strengths and weaknesses, as all of them offer unique advantages and limitations.
- X-ray crystallography and NMR spectroscopy are the two most common spectroscopic methods to determine the 3D structure of biomolecules at high resolution.
- X-ray crystallography requires the preparation of protein crystals; the obtained structure is a snapshot of the solid state. NMR can be used to determine biomolecular structures in aqueous solutions or the solid state.
- NMR has the edge over the X-ray method as it can be used to study biomolecules in their native state (*i.e.*, in solution) under physiological conditions or, as in the case of proteins, even in a molten globule or a denatured state.
- X-ray crystallography provides the structural information of a biomolecule as a static image. In contrast, NMR can provide information on the dynamic structure of biomolecules as the data is collected in the solution phase.
- The most important advantage of NMR spectroscopy is its ability to monitor dynamic events on a broad range of time scales, from fast (less than a nanosecond) to slow (milliseconds to seconds) internal motions

without perturbing chemical and structural equilibria. Thus, NMR spectroscopy can provide insights into atomic resolution concerning both structure and dynamics. Such combined structure-dynamic information facilitates the efficiency of structure-based design and accelerates the discovery of novel drugs.

- The most important advantage of X-ray crystallography is that it can determine the structure of proteins of any size and complexity, from low kilodaltons to even megadaltons (such as virus particles, membrane pores, ribosomes, *etc.*). NMR spectroscopy is limited to solving 3D structures of proteins with molecular masses less than ~30 kDa. It is difficult to determine the structure of a protein of more than 30 kDa using NMR due to the increased complexity and reduced sensitivity of an NMR spectrum with an increase in the size of the protein. Most structure-based drug discovery scientists use X-ray crystallography for their work.

Table 1: Comparison of the Advantages and Limitations of X-ray Crystallography and NMR Spectroscopy

Method	*Advantages*	*Disadvantages*
X-ray crystallography	Well-developed method. High resolution. No limitation on protein size. Easy model building steps. Crystallizable samples are soluble proteins, membrane proteins, ribosomes, DNA, RNA, and protein complexes.	Difficulties encountered during crystal preparation. Difficulties in diffraction. Solid structure is preferred. Static crystalline state structure is obtained.
NMR spectroscopy	High resolution. The 3D structure can be obtained in the solution state. Suitable for dynamics studies.	Requires highly purified sample. Complex sample preparation steps. Structure solution is challenging for large proteins. Suitable for proteins below 30 kDa. Requires water-soluble samples.

Chapter 9

Electron Spin Resonance Spectroscopy

1. Introduction

- Electron spin resonance (ESR) or electron paramagnetic resonance (EPR) spectroscopy is a powerful tool for studying materials with unpaired electrons (paramagnetic species). For *e.g.*, free radicals, transition metal ions, organic and inorganic radicals, and triplet states.
- The basic principle behind ESR is similar to NMR spectroscopy, *i.e.*, it is a magnetic resonance technique. However, ESR spectroscopy uses the interaction of an external magnetic field with the unpaired electron(s) rather than the nuclei of individual atoms (as in NMR spectroscopy).
- In ESR spectroscopy, the coupling between the unpaired electrons and magnetic nuclei is measured.
- The radiations with a frequency in the microwave region (0.04–25 cm) are absorbed by paramagnetic species resulting in transitions between magnetic energy levels of electrons having unpaired spins. Thus, ESR spectroscopy is a type of absorption spectroscopy.
- ESR spectroscopy is used to detect radicals and determine their concentration, structure, and chemical function.
- It is also used to study the kinetics, mechanisms, and structures of paramagnetic species. It has wide applications in the fields of biochemistry, structural biology, chemistry, medicine, geoscience, and polymer science.

2. Principle of ESR Spectroscopy

- ESR spectroscopy is based on the fact that any species with an odd number of electrons exhibit characteristic magnetic properties. An electron has a spin, due to which it acts as a tiny magnet and has a magnetic moment.
- When a molecule containing an unpaired electron is placed in a magnetic field, the spin of the unpaired electron can align in two ways to create two spin states, $m_s = \pm \frac{1}{2}$.
- The alignment can either be parallel (along the direction) to the applied magnetic field corresponding to the lower energy state, $m_s = -\frac{1}{2}$, or antiparallel (opposite) to the direction of the magnetic field, $m_s = +\frac{1}{2}$.
- The two alignments have different energies, and the energy difference between the two spin states is given by:

$$\Delta E = E_+ - E_- = h\nu = g\beta B$$

where, h is Planck's constant (6.626×10^{-34} J.s^{-1}), ν is the frequency of radiation, β is the Bohr magneton (9.274×10^{-24} J.T^{-1}), B is the strength of the magnetic field in Tesla, and g is the g-factor, the g-factor is a dimensionless measurement of the electron's intrinsic magnetic moment, and its value for a free electron is 2.0023.

- However, the value of g can vary and can be calculated from the above equation as follows:

$$g = \frac{h\nu}{\beta B}$$

- The spin state on the electron can flip when a light in the microwave range is applied.
- Since h, ν, and β are constant during an experiment, the values of g decrease as B increases.
- The concept of g can be roughly equated to that of chemical shift in NMR.

3. Construction of an ESR Spectrometer

- ESR spectroscopy can be carried out by two methods:
  1. Varying the magnetic field and holding the frequency constant, or
  2. Varying the frequency and holding the magnetic field constant (as an NMR spectrometer).
- Typical ESR spectrometers use the first approach and vary the magnetic field while holding the frequency constant (opposite to NMR spectrometers).
- The majority of ESR spectrometers are in the range of 8-10 GHz (X-band), though lower and higher fields ESR spectrometers are available, such

as 1-2 GHz (L-band), 2-4 GHz (S-band), 35 GHz (Q-band) and 95 GHz (W-band).

- An ESR spectrometer has the following major components: (1) Microwave generator (klystron), (2) attenuator, (3) circulator/isolator, (4) sample cavity, (5) two magnets, and (5) detector. **Figure 1** shows the components of a typical ESR spectrometer.

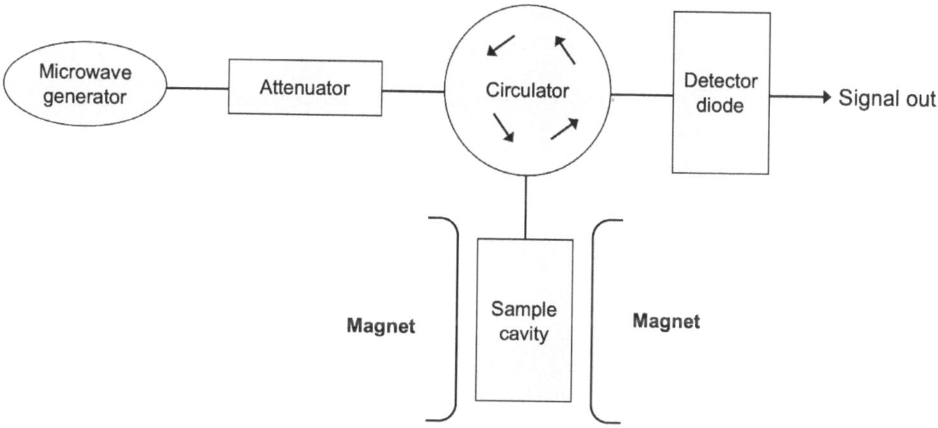

Figure 1: A Typical ESR Spectrometer.

- Any type of sample, *i.e.*, solid, powder, crystal, or solution, can be used for ESR spectroscopy.
- The klystron generates microwaves that pass through an attenuator and reach the sample. Microwaves reflected from the cavity are routed to the detector. The signal is received in the form of a decrease in current at the detector, analogous to the absorption of microwaves by the sample. The sample is kept in a sample cavity between the two poles of an electromagnet, and a stable and uniform magnetic field is applied over the sample volume. The ESR spectrum is recorded by steadily varying the magnetic field by sweeping the current supplied to the magnet.
- An ESR spectrum is reported as the first derivative of the absorption spectrum **(Figure 2)**.
- The unit of magnetic field strength is either Gauss or mTesla.
- To quantitate the ESR signal, double integration of the first derivative ESR spectrum is performed. Since the ESR signal is recorded as a first derivative of the absorption signal, the spectrum must be integrated once to recover the absorption spectrum and then integrated a second time to obtain the area under the absorption curve.
- ESR spectra are often complicated and analyzed using computer software that performs the double integration and predicts the ESR spectra.

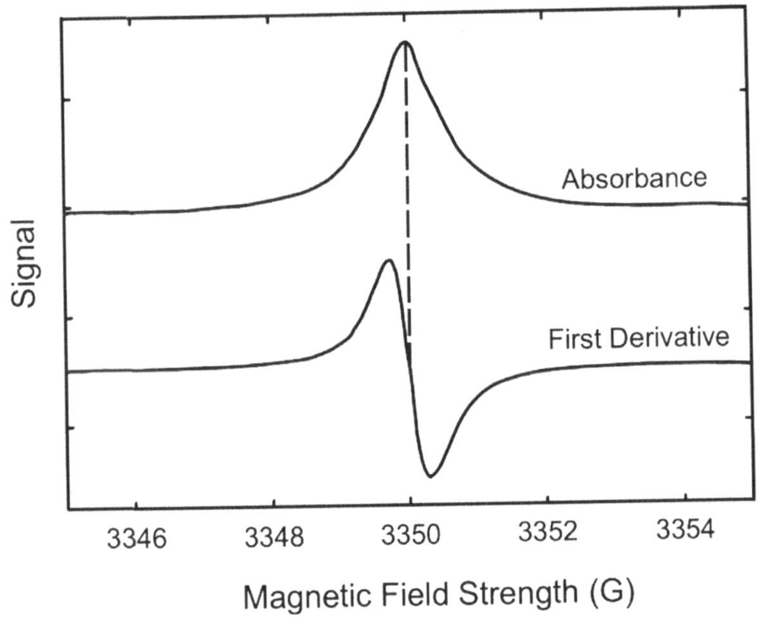

Figure 2: An Example of the First and Second Derivative ESR Spectrum.

4. Applications of ESR Spectroscopy

- ESR spectroscopy is the most common method to study transition metal-containing metalloproteins.
- The structure of cytosolic and membrane proteins and nucleic acids can be determined using spin labels and paramagnetic probes.
- ESR spectroscopy can be used to determine the rate of catalysis of different metals, as well as can study enzyme-ligand interaction.
- To understand the active site geometry of an enzyme.
- It is also used to study protein denaturation and folding/unfolding.
- *In vivo* ESR spectroscopy is used for quantitative and qualitative analyses of reactive oxygen species (ROS) and reactive nitrogen species (RNS) in tissues.
- The antioxidant levels in food and beverage samples can be quantified using ESR spectroscopy
- It is also used to study natural substances such as minerals with transition elements, minerals with defects (*e.g.*, quartz), petroleum, coal, rubber, *etc.*
- Specialized non-invasive ESR imaging can spatially locate paramagnetic species in animals. Such imaging is also used to determine the type, amount, and location of radicals in materials science.

Chapter 10

Atomic Absorption Spectroscopy

1. Introduction

- ☆ Atomic absorption spectroscopy is a technique to determine the concentration of a particular metal in a sample.
- ☆ Thus, it is used for the quantitative and qualitative determination of elements.
- ☆ The technique utilises absorption spectroscopy to assess the concentration of an element/analyte in a sample.
- ☆ It obeys the Beer-Lambert law.
- ☆ Atomic absorption spectroscopy is also known as flame atomic absorption spectroscopy since the method employs flame.

2. Principle of Atomic Absorption Spectroscopy

- ☆ It is based on the principle that free electrons generated in an atomiser absorb radiation of different wavelengths.
- ☆ The free electrons absorb UV or visible light. This cause the electrons to transfer to higher energy orbits. During this process, the absorption spectrum is produced, which is recorded by the photodetectors.
- ☆ The absorption spectrum is used to quantify free electrons in the gas state of the element.

- The amount of photon (radiation) absorbed results in an absorption spectrum that can then be measured as absorbance. The absorbance of a sample is directly proportional to the concentration of the element in the sample.
- A specific wavelength of light is absorbed by an atom depending on the nature of the element. For *e.g.*, sodium is absorbed in 589 nm, uranium is absorbed in 589 nm, and potassium is absorbed in 766.5 nm. The electronic transition is specific to a particular element. From the ground state, an atom is excited to an electronic state by absorption of energy of light at a particular wavelength.

3. Steps in AAS

- Atomic absorption spectroscopy is carried out in three steps:
  1. Conversion of the analyte molecule into its constituent gas-phase atoms (atomisation).
  2. Subsequent absorption of radiation by these free atoms.
  3. Measurement of the amount of absorbing species produced at a given wavelength.

4. Instrumentation

4.1. Atomiser

- Atomisation can be carried out either by a flame or furnace. Heat energy from flames, electrical discharges or plasma is used to convert metallic elements to atomic dissociated vapour. At very high temperatures, atoms can be ionised, so the temperature should be controlled carefully.
- To create a flame, fuel and oxidant are used. Usually, natural gas, hydrogen, propane, butane, and acetylene are used as fuels. While air, oxygen, or nitrous oxide are used as an oxidant.
- The fuel and oxidant gases are purged into a mixing chamber and then to the burner. The sample is aspirated into the mixing chamber.

4.2. Hollow Cathode Lamp

- A hollow cathode is used to glow a discharge lamp to produce sharp emission lines for a specific element.
- The metal used in the cathode is the same as the metal that is under investigation. The lamp is filled with an inert gas at low pressure. The lamp produces a glow of emission from the hollow cathode.

4.3. Monochromator

- A desired band of lines is isolated with a monochromator by passing a narrow band. The spectrum through a monochromator is shown as a curve.

4.4. Detector

- ☆ It converts light coming from the monochromator to a simplified electrical signal. Generally, a photomultiplier tube is used. A detector can be tuned to respond to a specific wavelength or frequency.

4.5. Recorder

- ☆ The recorder receives the electrical signals from the detector to convert them into a readable response. A computer system with suitable software is commonly used to record signals from the detector.

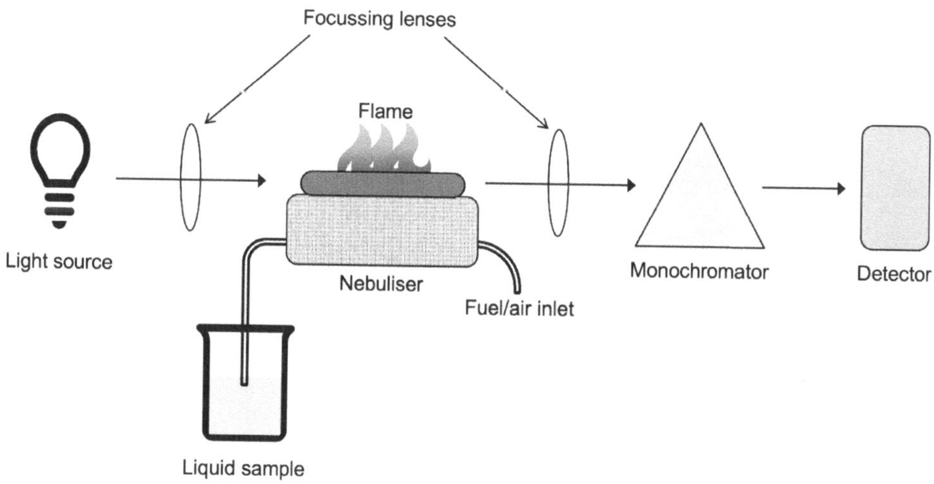

Figure 1: Schematic Representation of an Atomic Absorption Spectrometer. The main components in the instrumentation are an atomiser, a light source (hollow cathode lamp), a monochromator, and a detector (with recorder).

5. Experimental Procedures

Step 1: Atomisation

- ☆ Atomic absorption spectroscopy requires atoms to be in their gaseous phase. Therefore, the sample must be vaporised or desolved at high temperatures.
- ☆ The liquid sample is converted into atomic gas in three steps:
  1. *Desolvation:* The liquid solvent is evaporated, and the dry sample remains. This is done through nebulisation or spraying.
  2. *Vaporisation:* Solid sample is vaporised to a gaseous state.
  3. *Volatilisation:* Molecules making up the sample are broken into free atoms.

Step 2: Absorption of Radiation by the Free Atoms

- The electrons of the atoms in the atomiser absorb radiation of different wavelengths and are promoted to higher orbitals.
- The amount of energy (or wavelength) is specific to a particular electronic transition in a particular element, and each wavelength corresponds to only one element.

Step 3: Measurement of the Absorbing Species/Free Atoms

- As the amount of energy given into the flame is known, and the amount remaining can be measured at the detector, it is possible from the Beer-Lambert law to calculate how many transitions took place and get a signal.
- The signal is directly proportional to the concentration of the element being measured.
- A graph is plotted for the absorbance against the concentration of the element in the sample.

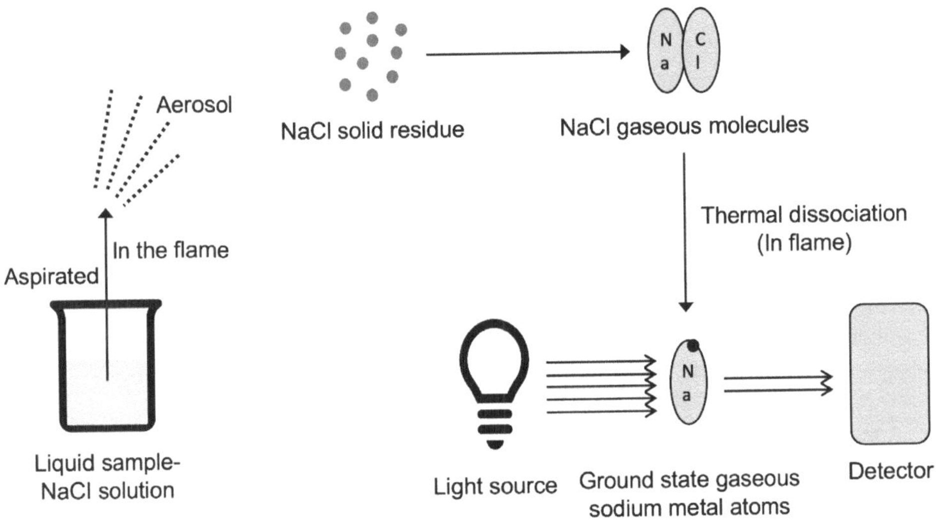

Figure 2: Steps in Atomic Absorption Spectroscopy.
Details are mentioned in the text.

6. Applications of AAS

- Atomic absorption spectroscopy is a powerful technique used in the qualitative and quantitative analysis of metallic elements in biological and chemical samples.
- It has several advantages, such as it can analyse multiple samples within a day and determine elements at trace concentrations. In addition, it does

not require pure elements before analysis as it can be used to distinguish one element in the presence of another.

- It has applications in analytical chemistry, forensics, and food industries.
- It is a common method to detect metals in pharmaceutical and oil products.
- It is also used in detecting metals as impurities in alloys and other mixtures.
- It is an important method for analysing environmental samples like water and soil.

Chapter 11

Turbidimetry and Nephelometry

1. Introduction

- ☆ When light strikes a particle in solution, some are absorbed by the particle, some are transmitted through the solution, and some are scattered or reflected. The amount of light transmitted or scattered is proportional to the concentration of insoluble particles.
- ☆ The intensity of transmitted light is measured using turbidimetry.
- ☆ The intensity of scattered light is measured using nephelometry.
- ☆ The turbidity in a liquid sample is caused by the presence of tiny suspended particles.
- ☆ The intensity of transmitted and scattered light is a function of the concentration of suspended particles.
- ☆ Thus, turbidimetry and nephelometry are spectroscopic techniques based on light scattering.
- ☆ These methods are used to determine the amount of turbidity in a solution based on the measurement of the transmission and scattering of light.
- ☆ These methods depend on the elastic scattering of radiation by a suspension of colloidal particles.
- ☆ To understand these methods, you must read the processes associated with light scattering (chapter 6 Raman scattering).

- Turbidimetry and nephelometry are useful for measuring precipitates formed by the interaction of very dilute solutions or other particulate matter, such as suspensions of bacterial cells. To obtain consistent results, all methodological variables must be carefully controlled. Birefringence-related problems might arise, particularly while working with bacterial cells. Highly diluted solutions are preferred.

Note

- The blue colour of the sky and the red colour of the sun at sunset result from the scattering of light from small dust particles and other gases in the atmosphere.
- The efficiency with which light is scattered depends on its wavelength, λ.
- The sky is blue because violet and blue light are scattered much more than other light wavelengths.

2. Turbidimetry

- In turbidimetry, the light is passed through a medium with dispersed particles, and the intensity of light transmitted is measured.
- Any standard calorimeter or spectrophotometer (visual or photoelectric) can be used as a turbidimeter.
- Photovoltaic cells and phototubes are suitable for turbidimetric analysis.
- The turbidometric measurements are made at 180° from the incident light beam.

3. Nephelometry

- In nephelometry, the intensity of scattered light is measured as a function of the concentration of the dispersed phase.
- Nephelometry is most sensitive for dilute suspensions.
- The nephelometric measurements are usually made at right angles to the incident light beam.

4. Factors Affecting Light Scattering

1. **Concentration of particles:** Light scattering obeys Beer-Lambert's law at low particle concentrations.
2. **Particle size:** The fraction of light scattered depends on the size and shape of particles. A larger particle size leads to more scattering and vice versa.
3. **Wavelength:** The intensity of scattered light depends on the wavelength of the incident light. Shorter wavelengths are scattered more than longer ones.

4. **Other factors:** Several other factors also influence the optical properties of suspension and particle size, such as temperature, rate of mixing of solutions, presence of extra ions or non-electrolytes, stability of dispersions, *etc.*

5. Applications of Turbidimetry and Nephelometry

5.1. To Study Air and Water Pollution

☆ Applications of nephelometry and turbidimetry are in the measurement of air and water quality (for pollution monitoring). In water, turbidity is monitored, whereas in air, smoke and dust are monitored. They have applications in climate monitoring, global warming (measuring global radiation levels), *etc.*

5.2. Inorganic Analysis

☆ Another use of nephelometry and turbidimetry is the determination of inorganic salts, such as sulfate as $BaSO_4$, chloride as AgCl, *etc.*

5.3. Biological Applications

☆ They are also used for the quantitative determination of proteins and other antigens in serum, urine, cerebrospinal fluid, *etc.*

Chapter 12

Mass Spectrometry

1. Introduction

- Every matter is made up of atoms; many atoms together make up molecules. Atoms consist of protons, neutrons, and electrons.
- How to calculate the weight of an atom? *i.e.*, how much does an atom weigh?
- As atoms and molecules are very small, it is not practical or convenient to measure their masses in kilograms or grams. For *e.g.*, the mass of a single hydrogen atom is approximately 0.00000000000000000000000166 grams (1.66×10^{-24} grams).
- Similarly, 12 grams of carbon contains 6.02×10^{23} atoms of carbon. Hence one atom of carbon weighs 1.99×10^{-23} grams.
- To overcome the difficulties of using conventional measurement units, scientists use the unit of mass called the Dalton (Da). Da is a convenient unit for studying the mass of molecules. It is defined as 1 Da = (1/12) of the mass of a single atom of the isotope of carbon-12 (C^{12}).
- Mass spectrometry is a method to weigh these tiny masses. It is an analytical tool used to measure the molecular mass of a sample. It is carried out in a mass spectrometer.
- Mass spectrometry involves molecules that are analyzed by their fragmentation patterns, not by their interaction with electromagnetic radiation. Thus, the use of the term mass spectroscopy is perhaps incorrect.

- ☆ The underlying principle of mass spectrometry is straightforward. A moving charged particle can be deflected by applying electric and magnetic fields. The deflection depends on the mass and the charge of the particle.
- ☆ The mass spectrometer determines the mass of a molecule by measuring the mass (m) to charge (z) ratios (m/z) of ions.
- ☆ Structural information can also be generated using certain types of mass spectrometers, such as tandem mass spectrometers. This method is helpful for the structural elucidation of organic compounds and peptide or oligonucleotide sequencing.

Note

- ☆ By definition, mass spectrometry is an analytical technique used to quantify known materials, identify unknown compounds within a sample, and elucidate the structure and chemical properties of molecules.
- ☆ **Why is ionization necessary?:** It is difficult to manipulate neutral particles. If they are charged, then electric and magnetic fields can be used to move them around.

- ☆ The complete process of mass spectrometry involves converting the sample into gaseous ions, with or without fragmentation. These ions enter the analyzer region of the mass spectrometer, where they are separated according to their m/z ratios. The separated ions are detected, and the signal is sent to the recorder, where the m/z ratios are stored with their relative abundance in the form of an m/z spectrum.
- ☆ The mass spectrometer separates ions by using a magnetic or an electrical field.
- ☆ Mass spectrometry can investigate any material that can be ionized and whose ions can exist in the gas phase. These include proteins, oligosaccharides, nucleic acids, and small molecules.
- ☆ Thus, a mass spectrometer generates multiple ions from the sample under investigation, then separates them according to their specific m/z ratio, and then records the relative abundance of each ion type

2. Principle of Mass Spectrometry

- ☆ The basic requirement for an analyte molecule to be studied using mass spectrometry is that it has to be charged. However, a few molecules may not be charged. Therefore, the first step in an MS experiment is to ionize the molecules. The spectrometer, therefore, has an ionization source. The ions generated are then separated by one or more mass analyzers which are then detected by a detector.

- ☆ The sample is passed to a vacuum chamber through an inlet. Depending on the type of inlet and ionization techniques used, the sample may already exist as ions in solution or be ionized in conjunction with its volatilization or by other methods in the ion source.
- ☆ **Figure 1** shows the components of a mass spectrometer and the overall workflow to obtain a mass spectrum.

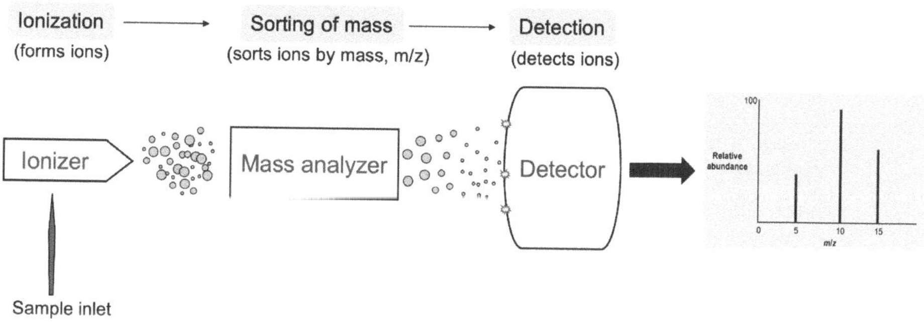

Figure 1: Basic Components of a Mass Spectrometer.
The overall workflow to obtain a mass spectrum is also shown.

- ☆ High-speed electrons knock off the electrons from the sample and change them to positive ions. An electric field then accelerates these ions through a magnetic field, which deflects the paths of the ions. The amount of path deflection for each ion depends on its mass (heaviest ions are deflected most and vice versa). Thus, the ions are separated.
- ☆ The ions are sorted in the mass spectrometer according to their m/z ratio, detected in proportion to their abundance, and then collected by a detector. In the detector, the ion flux is converted to an electrical current. The recorder detects the magnitude of the electrical signals as a function of m/z and converts it into a mass spectrum.
- ☆ A mass spectrum is a plot of the abundance of ions versus their m/z ratio. By comparing the positions where the ions hit the detector plate, accurate values of their relative masses can be determined. The mass spectrum is used to analyze the elemental or isotopic signature of a sample, the masses, and the chemical structures of molecules and compounds.
- ☆ In the mass spectrum of a pure compound, the molecular ion is present at the highest m/z value and gives the molecular mass of the compound. The position of the molecular ion is followed by ions with heavier isotopes.
- ☆ Mass spectrometry is handy for the detection and analysis of traces of macromolecules down to less than one pg (10^{-12} g).
- ☆ For example, when ^{13}C and ^{12}C are analyzed in a mass spectrometer, the ratio of their masses is found to be

$$\frac{\text{Mass } C^{13}}{\text{Mass } C^{12}} = 1.0836129$$

☆ Since the atomic mass unit is the mass of ^{12}C is exactly 12 atomic mass units, and thus on this same scale,

$$\text{Mass of } C^{13} = 1.0836129 \times 12 \text{ amu} = 13.003355 \text{ amu}$$

☆ Similarly, the mass of other atoms can be determined.

☆ In mass spectrometry, the mass-to-charge (m/z) ratios of ions and not the actual mass is measured.

☆ If, *e.g.*, a biomolecule is ionized by the addition of one or more protons (H^+ ions), the instrument measures the m/z after the addition of 1 Da for each proton if the instrument is measuring positive ions, or m/z minus 1 Da for each proton lost if measuring negative ions (**Figure 2**).

+1 charge state: $[M + H]$
+2 charge state: $[M + 2H]^{++}$
+3 charge state: $[M + 3H]^{+++}$

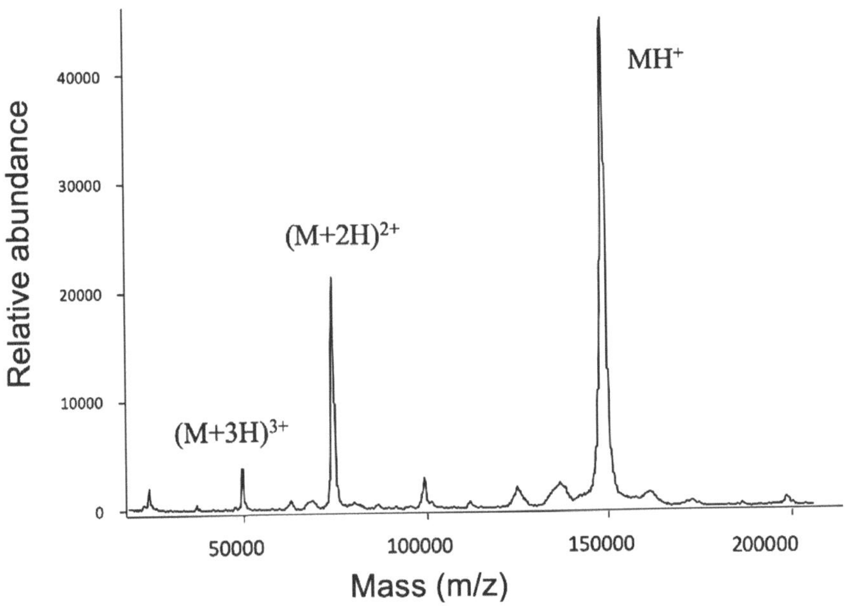

Figure 2: A Typical Mass Spectrum.

A mass spectrum is simply the m/z ratios of the ions present in a sample plotted against their relative abundance (intensities). Each peak shows a component of unique m/z in the sample, and the height of the peaks indicates the relative abundance of the various components in the sample.

☆ Most elements have more than one stable isotope. For *e.g.*, most carbon atoms have a mass of 12 Da, but in nature, ~1 % of C atoms have an extra neutron, making their mass 13 Da. This is important as the mass spectrometers can distinguish isotope peaks if their resolution is high enough. Thus, better mass accuracy is achieved in such mass spectrometers. The mass spectrum of a peptide with 94 C-atoms (19 amino acid residues) is shown in **Figure 3**.

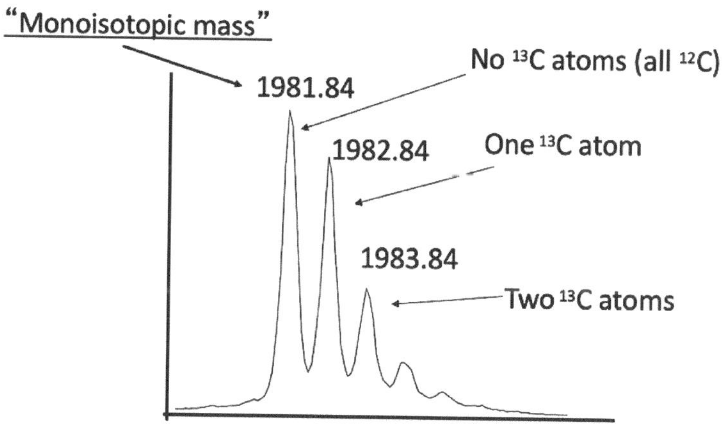

Figure 3: Mass Spectrum of a Peptide with 94 C-atoms (19 amino acid residues). When the isotopes are resolved, the monoisotopic mass is used as it is the most accurate measurement.

3 Construction of a Mass Spectrometer

☆ The instrument consists of three major components: (1) Ion source, (2) mass analyzer, and (3) detector. **Figure 4** shows the various components of a mass spectrometer and their types.

3.1. Ion Source

☆ The molecules of an analyte are converted into ions by the process of ionization.

3.2. Mass Analyzer

☆ Mass analyzers use electric and magnetic fields to accelerate ions and manipulate their flight in the gas phase, *i.e.*, separation of ions in the gas phase according to their *m/z* ratio.

3.3. Detector

☆ The detector detects the ions based on their *m/z* ratio and records the relative abundance of particular ions.

- Besides, a sample inlet system is used to inject the samples to be studied to the ion source while maintaining the high vacuum requirements (~10^{-6} to 10^{-8} mm of mercury) of the technique; and a computer to control the instrument, acquire and manage data, and compare spectra to reference libraries.
- To prevent collisions with gas molecules, the sample ionizer, mass analyzer, and ion detector are generally operated in a vacuum.
- The currently available ion sources, mass analyzers, and detectors are shown in **Figure 4**.

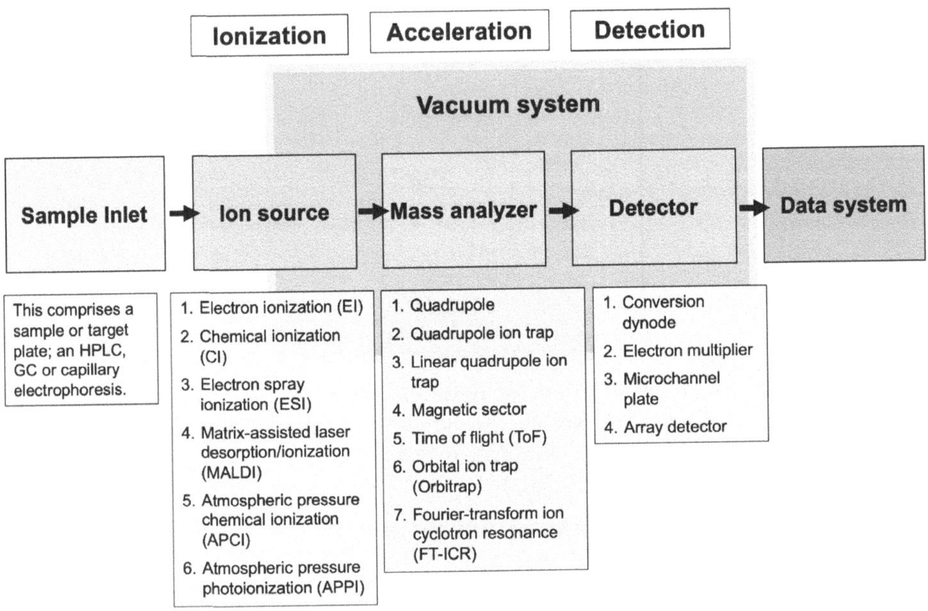

Figure 4: Various Components of a Mass Spectrometer and their Types. The details are discussed in the main text.

4. Types of Mass Spectrometers

4.1. Based on different Ion Sources

- The sample is inserted directly into the ionization source or passed through a chromatography instrument before entering the ionization source. When chromatography is used, the sample introduction usually involves the mass spectrometer coupled to gas chromatography (GC), high-pressure liquid chromatography (HPLC), or capillary electrophoresis (CE) columns. Hence, the sample is separated into a series of components that enter the mass spectrometer sequentially for individual analysis.

☆ There are several ionization methods in mass spectrometry. The ionization method depends on the sample type and the available mass spectrometer (**Table 1**).

Table 1: Various Ionization Methods.

Types of Samples	*Ion Source*
Gas	Electron ionization (EI), chemical ionization (CI), or photoionization (PI)
Solution	Electrospray ionization, atmospheric pressure chemical ionization, or atmospheric pressure photoionization
Solid	Matrix-assisted laser desorption/ionization (MALDI), plasma desorption, or fast atom bombardment

☆ Ions can be produced from a neutral molecule by adding an electron (to form an anion) or removing an electron (to create a positively charged cation).

☆ Both positive and negative-ion mass spectrometry may be carried out, but positive-ion mass spectrometry is more common. However, the principles of separation and detection are essentially the same for both types of ions.

4.1.1. Electron Ionization

☆ Electron ionization (EI) works well for gas-phase molecules and is the most common mass detector for gas chromatography (GC/MS).

☆ The electron beam is produced by a rhenium or tungsten filament at a temperature over 1000 K (**Figure 5**). The generated electrons are accelerated to 70 eV, forming an electron beam. The volatile sample or sample in the gaseous phase containing neutral molecules is introduced to the ion source perpendicularly to the electron beam. Interaction of electrons with the sample results in either losing an electron from the sample or adding an electron.

$$M + e^- = M^{\bullet +} + 2e^-$$

where, $M^{\bullet +}$ is the molecular ion. The loss of electrons is a negligible loss of mass; therefore, the mass of $M^{\bullet +}$ is equal to the mass of the molecule.

☆ Electron ionization results in extensive fragmentation, and molecular ions may not be observed for many compounds. However, fragmentation mass spectra are sometimes helpful as they provide structural information about a molecule.

☆ The chemical bonds in molecules are formed by electron pairing. An electron's impact may knock out one of the paired electrons, resulting in a bond with a single unpaired electron. This generates a radical that is also a

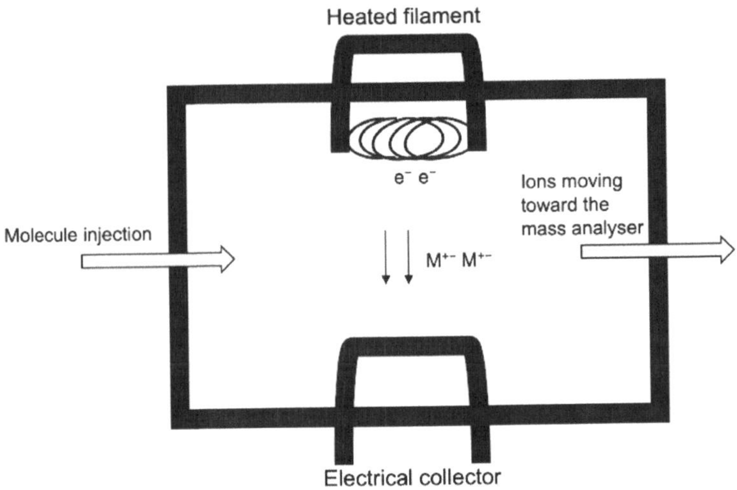

Figure 5: The Mechanism of Ion Formation in Electron Ionization.

cation, written as $M^{\bullet+}$, where (+) indicates the ionic state and (•) indicates the radical. Similarly, electron impact may result in electron capture (extra unpaired electron). This generates a radical that is also an anion, written as $M^{\bullet-}$, where (-) indicates the ionic state and (•) indicates radical. Such radical ions are called molecular ions, parent ions, or precursor ions and are relatively unstable. Their energy has to be dissipated.

- ☆ Upon dissipation of energy, the precursor ion disintegrates into several smaller fragment ions that are relatively unstable and undergo further fragmentation. This produces a series of daughter ions or product ions that is recorded as a mass spectrum.
- ☆ The electron ionization spectrum of n-butane ($CH_3CH_2CH_2CH_3$) is shown in **Figure 6**.

4.1.2. Chemical Ionization

- ☆ The ions are produced by colliding the sample molecules with the primary ions produced by a gas (called a reagent gas) in the ionization chamber. The reagent gas is ionized through electron ionization. The radical cations generated will undergo fragmentations and reactions.
- ☆ Chemical ionization (CI) is useful for determining molecular mass, as high-intensity molecular ions are produced due to less fragmentation.
- ☆ However, only a limited amount of structural information can be determined about the ionized species.
- ☆ Chemical ionization requires less energy than electron ionization (soft ionization), resulting in lesser fragmentation than the electron ionization method.

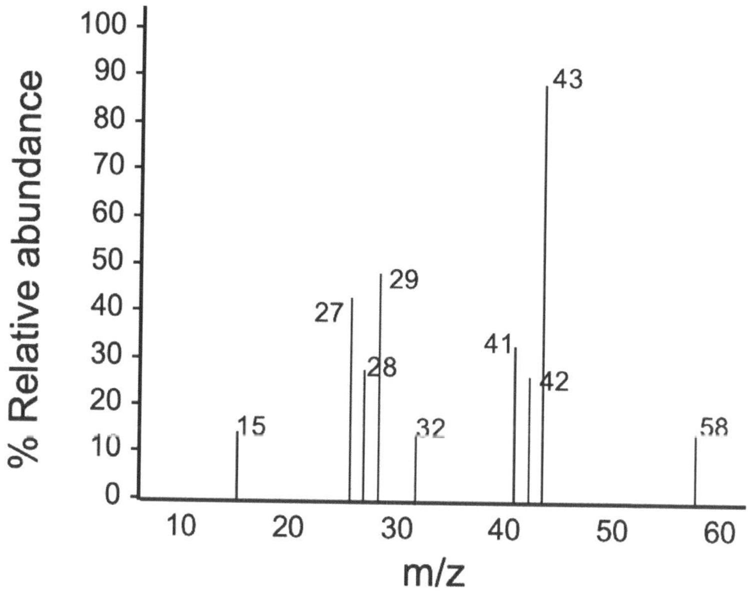

Figure 6: The Electron Ionization Spectrum of n-butane.

The relative abundance is plotted from 0 to 100 % in the spectrum, where the most prominent peak is set at 100 % (base peak).

- ☆ The setup for chemical ionization is similar to electron ionization. However, in this method, a gas like CH_4 or NH_3 is injected into the ion chamber. Due to electron impact, the gas ionizes. This follows the injection of the analyte molecules. The analyte molecules undergo collisions with the gas; the gas ions in this cloud react with the analyte to produce analyte ions that are excellent proton donors for the analyte.
- ☆ Like electron ionization, chemical ionization is suitable only for gaseous samples (gases and volatile liquids).

> ***Note***
>
> ☆ **Soft ionization:** ionization of molecules into gas-phase ions. Too much energy (hard ionization) breaks molecules into much smaller fragments.

4.1.3. Electrospray Ionization

- ☆ Electrospray ionization (ESI) is a preferred ionization method when the sample is in liquid form.
- ☆ This is also a soft ionization technique as the sample is ionized by adding or removing a proton, with minimal energy remaining to cause fragmentation of the sample ions. This results in minimum fragmentation of the sample. Thus, it is a very important method for analyzing biomolecules, such as proteins and DNA.

- ☆ The analyte solution is introduced either from a syringe pump or as the eluent flow from liquid chromatography **(Figure 7)**. It then passes through the electrospray needle (capillary with 75-150 µm internal diameters) with high potential (2.5-4 keV). Electrospray ionization results from the strong electric field (1-4 keV) acting on the surface of the sample solution. This forces the spraying of charged droplets from the needle with a surface charge of the same polarity as the charge on the needle. As the droplet moves towards the counter electrode (which passes it to the analyzer), solvent evaporation occurs, and the droplet shrinks until it reaches the point that the surface tension can no longer sustain the charge (the Rayleigh limit), and at that point, droplets break. This produces smaller droplets, and the process is repeated. As the solvent evaporates, the droplet size decreases, and eventually, only the charged analyte (free of solvent) remains.

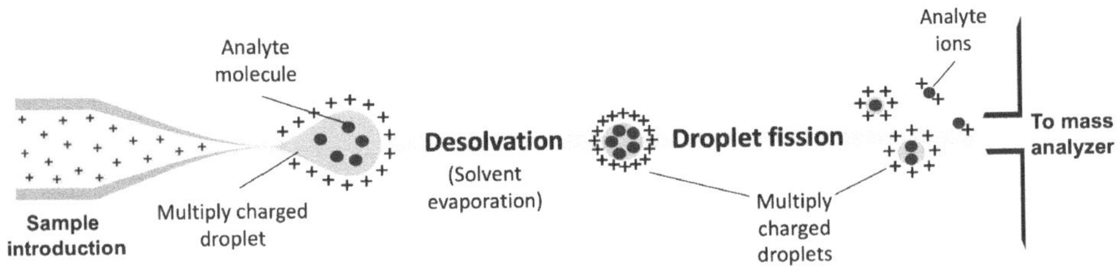

Figure 7: The Process of Electrospray Ionization.

The sample enters from the capillary towards the MS inlet cone. Due to desolvation, the charged solvent droplet shrinks until the surface tension holding the charged droplet is overcome by Coulombic repulsion. This results in droplet fission and the generation of gas-phase analyte ions.

- ☆ These charged analyte molecules can have single or multiple charges depending on the availability of ionizable groups. Biomolecules usually result in mass spectra wherein the consecutive peaks differ by one charge unit. This information allows one to calculate the mass of the biomolecule.

Note

- ☆ Nanospray ionization is a low flow rate version of electrospray ionization. A small volume (1-4 µL) of the sample dissolved in a solvent is transferred into a miniature sample vial.
- ☆ Ionization can also occur at atmospheric pressure; this method is called atmospheric pressure ionization (API).
- ☆ In field desorption (FD), the sample deposited on the filament is gradually heated by the electric field. Sample ionize by electron tunnelling.

- In field ionization (FI), the sample introduced in the gas phase (heated or not) is ionized by electron tunnelling near the emitter.
- In fast atom bombardment (FAB), the analyte is in a liquid matrix (such as glycerol, *etc.*). The bombardment is with a fast atom beam (xenon at 6 keV). The desorption of molecular ions and fragmentation occurs, and matrix clusters.

4.1.4. Matrix-Assisted Laser Desorption Ionization

- Matrix-assisted laser desorption ionization (MALDI) is also a soft ionization technique and results in minimum fragmentation of the sample.
- MALDI is used for solid, non-volatile, and thermally labile compounds such as proteins, oligonucleotides, bio- or synthetic polymers, *etc.*
- The sample is mixed in a solution containing small organic molecules called the matrix. The sample is then spotted onto a metal plate and dried, where it co-crystallizes with the matrix on drying.
- Commonly used matrices are α-cyano-4-hydroxycinnamic acid (for protein, peptides, organic compounds, <10 kDa), 3,5-Dimethoxy-4-hydroxycinnamic acid (for proteins >10 kDa), and 2,5 dihydroxy benzoic acid (for oligonucleotides, synthetic polymers carbohydrates).
- The next step occurs inside the source of the mass spectrometer under a vacuum. Intense laser pulses are bombarded on the dried sample for a short duration, which causes rapid excitation and vaporization of the crystalline matrix. The matrix absorbs the laser energy causing a small part of the analyte ions to convert into the gas phase that is analyzed in a mass analyzer (generally a time-of-flight mass analyzer, TOF-MS). However, the exact process of forming analyte ions is still unclear. The ionization process in MALDI is shown in **Figure 8**.

Note

- MALDI and ESI are soft ionization methods. The peptides/proteins do not fragment during these processes; thus, they are helpful for molecular weight determination. However, sometimes fragmentation is useful because it provides valuable information about the amino acid sequence of peptides.
- The matrix is dissolved in solvents like acetonitrile or trifluoroacetic acid.
- The matrix compound contains ring structures that absorb energy from UV lasers. It also has acidic groups to protonate peptides.
- TOF is the best type of mass analyzer to couple to MALDI, as TOF has a virtually unlimited mass range.

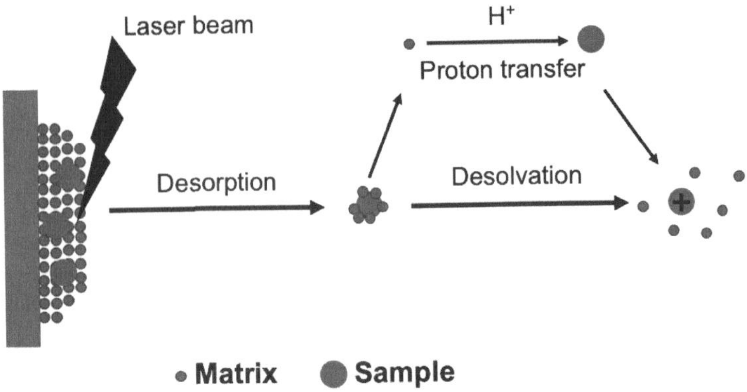

Figure 8: Ionization Process in MALDI.

The sample mixed in a matrix is placed on a target and introduced into a high vacuum chamber. Desorption and ionization are achieved by applying the laser pulse to the dried sample. Although the exact process of forming analyte ions is still unclear, it is believed that the absorption of light by the matrix molecules causes the sublimation of matrix crystals with the analyte molecules into the gas phase.

4.2. Based on different Mass Analyzers

- ✰ Once ions are created and leave the ion source, they enter a mass analyzer that separates them and measures their masses (actually, the mass-to-charge (m/z) ratio).
- ✰ At a certain moment, ions of a specific mass are allowed to pass through the analyzer and are counted by the detector. Afterwards, ions of a different mass are allowed to pass through the analyzer, and again the detector counts the number of ions. In this manner, the analyzer scans through a wide range of masses.
- ✰ In a mass spectrometer instrument, a particular type of ionization method is coupled to a particular mass analyzer that operates on a specific principle. *i.e.*, EI, CI, and FAB are combined with magnetic sector instruments, ESI and its derivatives are coupled to quadrupole (or its variant ion-trap), while MALDI is coupled to TOF detection.

4.2.1. Quadrupole Mass Analyzer

- ✰ Mass analysis in a quadrupole mass analyzer (Q-MS) is based on obtaining a stable trajectory for ions of specific m/z values in a rapidly changing electric field.
- ✰ As the name implies, the quadrupole mass analyzer consists of 4 parallel cylindrical rods **(Figure 9)**. To one pair of diagonally opposite rods, a negative DC (-dc) voltage and an alternative radio frequency (RF) voltage are applied, while to the other pair of rods, a positive DC voltage (+dc)

of opposite polarity and the inverse RF voltage are applied. This creates a continuously varying electric field along the length of the analyzer.

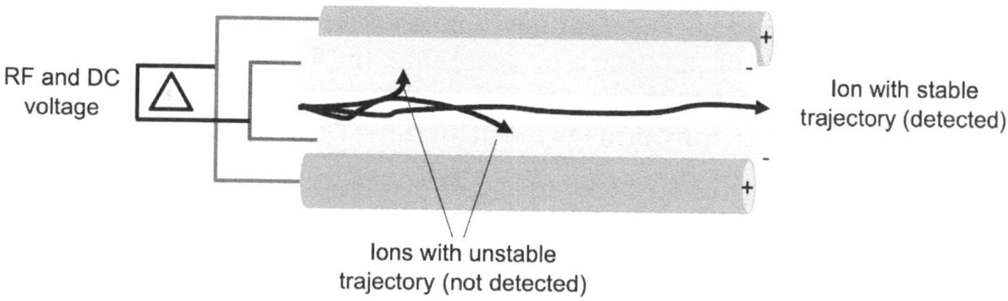

Figure 9: Quadrupole Mass Analyzer.

Ions pass through four parallel metal rods of opposing charges. The application of an RF voltage with a DC offset causes ions to oscillate. Ions with an unstable trajectory will crash into the metal rods, filtering ions with stable trajectories to pass through to the next sector of the instrument.

- ☆ Ions oscillate inside the quadruple as a result of the high-frequency electric field. The fixed (DC) and oscillating (RF) fields cause the ions to undergo complicated trajectories through the quadrupole filter. Ions are accelerated, and mass separation occurs. At a given DC/RF voltage, a specific mass/charge ion reaches the detector.
- ☆ By changing the electric field (scanning), the ions of different *m/z* successively arrive at the detector.
- ☆ When an unknown sample is analyzed, the DC/RF voltage is varied to get the signal at the detector. Once a DC and RF voltage combination gives a signal, *m/z* can be calculated.

Note

- ☆ Ion trap mass analyzers are very similar to quadrupoles. However, they consist of a ring electrode and two end caps, and the ions are stored by RF and DC fields.

4.2.2. Time-of-Flight Mass Analyzer

- ☆ In a time of flight (TOF) mass analyzer, the time taken by the ions to reach the detector is measured. The ions are generated in bundles, *e.g.*, by MALDI. The ions are then accelerated towards the flight tube, and in the absence of any electric field, the ions drift according to their velocities.
- ☆ Mass analysis in a time-of-flight mass analyzer (TOF-MS) is based on the fact that ions of different *m/z* values have the same energies but different velocities after acceleration from the ion source.

- The time required for each ion to pass the tube differs for different ions: ions with lower mass ions reach the detector quicker than those with higher mass. An *in-built* clock is used to measure the time of flight.
- A TOF mass analyzer separates ions based on their velocity. The ions are accelerated by the electric field and enter the flight tube. The lighter ions travel faster than the heavier ions to the detector (separated according to mass as they fly through the flight tube) **(Figure 10)**.

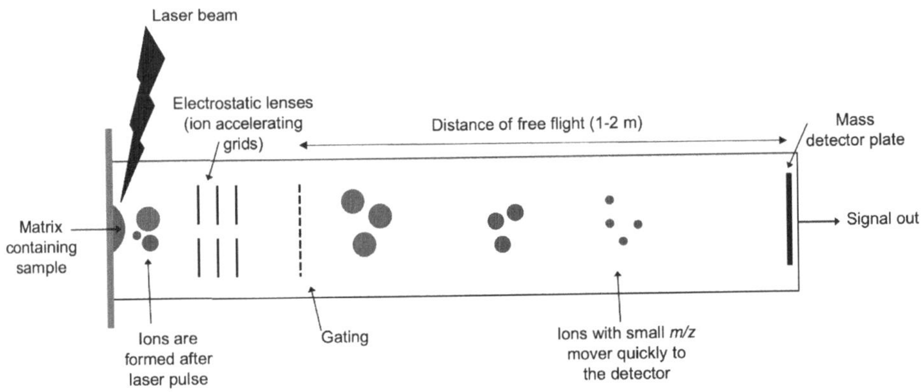

Figure 10: Time-of-Flight (TOF) Analyzer.

The ions are accelerated by an electric field with the same electrical potential, and then the time taken by each ion to reach the detector is measured. If all ions have the same charge, their kinetic energies will be identical; thus, the velocity of each ion will depend only on its mass. Therefore, the lighter ions will reach the detector first, while heavier ions later.

- If the ions are accelerated with the same potential at a fixed point and a fixed initial time, they will separate according to their mass-to-charge ratios. This time of flight can be converted to mass.
- Ions strike the detector either in linear or reflectron mode at different times, depending on their m/z ratio.
- A data system controls instrument parameters, acquires signal versus time, and processes the data.
- Typically, a few 100 pulses of laser light are used, each of a few nanoseconds' duration, and the information is accumulated to build up a good spectrum.
- In a MALDI-TOF instrument, the sample mixed with a matrix is dried on the target plate and introduced into a high-vacuum chamber. A laser beam is focused on the sample using a camera, and the sample is irradiated with laser pulses.

4.2.3. Tandem Mass Spectrometry

- A tandem mass spectrometer (also known as MS/MS) is a mass spectrometer that has more than one analyzer; in practice, usually two (*i.e.*, have two mass analyzers combined in one instrument). Incorporating more than one mass analyzer dramatically enhances the sensitivity, mass resolution, and mass accuracy of the spectrometer.

- The two analyzers are separated by a collision cell. An inert gas (*e.g.*, argon or xenon) is injected into the collision cell to collide with the selected sample ions and fragment them. The process of fragmentation by collision is called collision-induced dissociation (CID).
- In simple words, in MS/MS, one set of ions (with one *m/z* value), called precursor ions, is selected from a mixture of ions; These ions are fragmented, and then the *m/z* of the resulting product ions are again measured **(Figure 11)**.

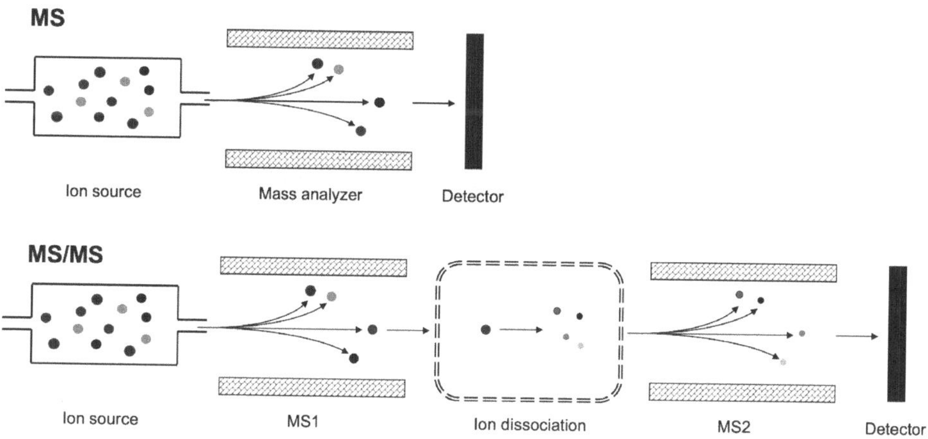

Figure 11: Comparison of MS (Top) and MS/MS (Bottom).

In MS, ions formed in the ion source are separated in the mass analyzer before reaching the detector. MS/MS is a two-step method where a specific ion is selected in the first mass analyzer (MS1) and then subjected to collision in the collision cell, whereas the second analyzer (MS2) separates the product ions before they reach the detector.

- The most common MS/MS experiment involves selecting an ion using the first mass analyzer, which is then fragmented into daughter ions in a collision cell. The daughter ions are then detected by a second analyzer. It is possible to further fragment the daughter ions into granddaughter ions that are then analyzed by a third mass analyzer.
- The use of multiple fragmentation steps can provide additional information about the structure of the analyte.
- MS/MS is widely used to elucidate the structural details of peptides, proteins, biomolecules, and other organic compounds.
- A common example of tandem MS instrumentation is the MALDI TOF/TOF or MALDI TOF/TOF tandem MS. In this instrument, two TOF analyzers are

separated by a pressurized collision cell, which fragments the precursor ion and then reaccelerates the product ions before they enter the second TOF analyzer. This instrument has the high mass accuracy associated with TOF mass analyzers, as well as the ability to do CID.

- MALDI-TOF/TOF is a rapid, accurate, and sensitive technique that can provide information on the molecular weights of proteins, peptides, oligonucleotides, nucleic acids, and carbohydrates.

Note

- MS/MS instruments with two, three, and four analyzers are commercially available.
- The analyzers do not necessarily have to be of the same type; the instrument may be a hybrid one.
- More popular tandem mass spectrometers are quadrupole-quadrupole, magnetic sector-quadrupole, and quadrupole-time-of-flight geometries.
- However, fragmentation can also be performed on certain single analyzer mass spectrometers such as ion-trap and time-of-flight instruments.

4.3. Based on different Detectors

- Ions from the mass analyzer impinge on a detector surface where the charge is neutralized, either by collecting or donating electrons.
- The detector monitors the ion current, amplifies it, and the signal is then processed by a computer, which records the data in the form of a mass spectrum.
- The m/z values of the ions are plotted against their relative abundance to show the components in the sample, their molecular mass, and their relative abundance in the sample.
- The type of detector used is based on the type of analyzer; the common detectors include photomultipliers, electron multipliers, and micro-channel plate detectors.
- The total ion current (TIC) is the sum of the current carried by all the ions being detected at any given moment and is a valuable parameter to measure during MS.

5. Mass Spectrometry Coupled with Chromatography

- Mass spectrometry is often combined with liquid and gas chromatographic methods, such as HPLC, FPLC, and GC. A schematic representation of the GC-MS instrument is shown in **Figure 12**.

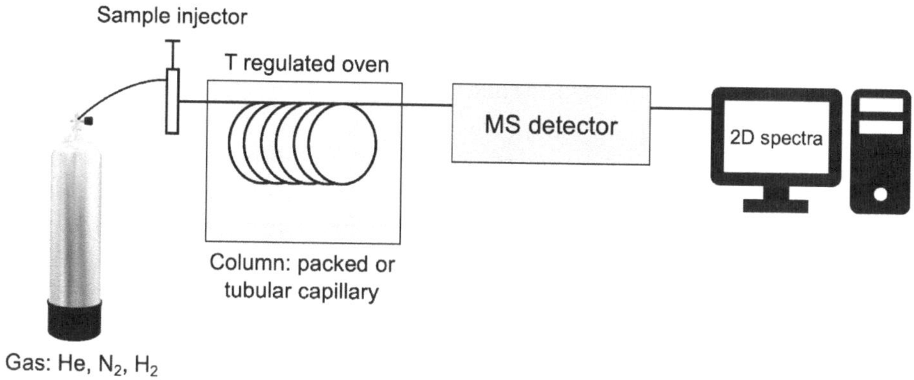

Figure 12: A GC-MS Instrument.

A GC/MS separates chemical mixtures (the GC component) and identifies the components at a molecular level (the MS component). An inert carrier gas transports the compounds through the column and to the detector. As the separated compounds emerge from the column opening, they flow into the MS. Mass spectrometry identifies compounds by their mass.

- ☆ Chromatographic methods separate the compounds based on the differences in their physicochemical properties, and mass spectrometry identifies compounds by their mass.
- ☆ The sample inlet may be coupled to a liquid chromatography system (LC-MS), gas chromatography system (GC-MS), or a capillary electrophoresis system.
- ☆ Two-dimensional spectra are obtained through the combination of MS with these methods. This dramatically enhances the analyzability of complex mixtures with a large number of components.
- ☆ For *e.g.*, ion-exchange chromatography on a cell extract with a resolution of 100 combined with MS with a resolution of 10,000 can result in a total resolution of almost 1,000,000 cellular proteins, where both methods are largely independent of each other.
- ☆ The EI or CI is used with the GC method (generally used for non-polar compounds), while ESI or MALDI is used with the LC method (generally used for polar compounds).

6. Protein Sequence Identification by Mass Spectrometry

- ☆ In recent years, mass spectrometry-based proteomics techniques have revolutionized the identification of proteins in biological samples. This has been made possible by improvements in sample preparation, instrumentation, and data analysis methodologies.

- Instead of sequencing entire proteins (top-down approach), which is laborious, bottom-up protein analysis uses mass spectrometry to identify peptides rather than proteins.
- In a bottom-up proteomics workflow, the protein sample is digested by a protease (most commonly trypsin), and the peptides are analyzed using LC-MS/MS. Peptides are separated and eluted from an LC column, and then peptides are ionized and transferred to the mass spectrometer, where their m/z ratio and relative intensities are measured. Whole peptide ions are subjected to fragmentation, generating a predictable series of fragment ions. The experimentally determined masses of intact and fragment ions observed by LC-MS/MS are compared with theoretical masses generated from a database of query protein sequences.
- The most common use of MS/MS in biochemical research is the product or daughter ion scanning experiments. This technique is highly successful for the sequencing of peptides and nucleotides.
- The detailed steps for protein identification using MS/MS are as follows:
  1. The protein of interest is digested with a suitable protease such as trypsin.
  2. The digested peptide mixture is analyzed by mass spectrometry directly, as an entire reaction mix, without separation or clean-up. A complex spectrum is obtained from which the molecular weights of all proteolytic fragments can be identified. With the molecular weight information, this spectrum is called a peptide map (if the protein sequence is available on a database, then the peptide map is generally sufficient to confirm the protein). For this experiment, the mass spectrometer is operated only in the MS mode, whereby the sample is sprayed and ionized, and the ions pass through the first (quadrupole) analyzer. The quadrupole, in this case, is not used as an analyzer; it is just used as a lens to focus the ion beam into the second analyzer (TOF), which separates the ions according to their mass-to-charge ratio.
  3. With the digested mixture still spraying into the mass spectrometer, the Q-ToF mass spectrometer is switched into MS/MS mode **(Figure 13)**. The protonated molecular ions of each digested fragment can be selected and transmitted through the quadrupole analyzer, which is used as an analyzer to transmit only the ions of interest into the collision cell. An inert gas is introduced into the collision cell, and the sample ions are bombarded by the collision gas molecules, causing them to fragment. The fragment (or daughter or product) ions are then analyzed by the second (time-of-flight) analyzer. An MS/MS spectrum is thus produced that shows all the fragment ions formed directly from the chosen precursor (parent) ions for a given peptide component.

4. The MS/MS fragment ion spectrum is produced for each component identified in the proteolytic digest. The spectra should be interpreted carefully, as varying amounts of sequence information can be obtained from each fragmentation spectrum. Some processing can be automated, but for reliable data, the processing and interpretation of spectra take time.
5 The amount of sequence information that can be obtained may vary. Some peptides can generate enough information for a full sequence to be determined; others may generate a partial sequence of a few (4-5) amino acids.

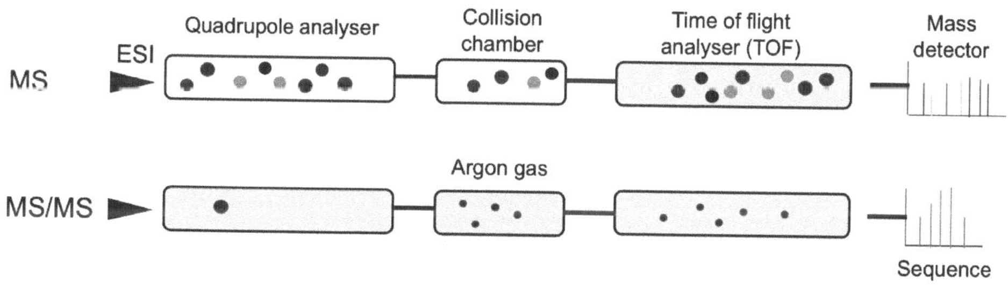

Figure 13: Q-TOF Mass Spectrometer Operating in MS (upper) and MS/MS (lower) Modes.

MS mode uses the first and second quadrupoles to provide an accurate mass scan of the unfragmented precursor ion. The MS/MS utilize mass filter mode to transmit ions into the collision cell where CID occurs. The subsequent product ions and any unfragmented precursor ions are then transmitted to the TOF analyzer, where accurate mass is measured.

7. Applications of Mass Spectrometry

1. In the structural analysis of biomolecules such as proteins, peptides, oligonucleotides, nucleic acids, and carbohydrates.
2. In determining the molecular mass of peptides, proteins, and oligonucleotides.
3. In sequence analysis of biopolymers such as proteins and oligosaccharides
4. In pharmaceutical analysis, such as drug metabolism, pharmacokinetics, characterization of drugs, analysis of drug degradation products, screening drug candidates, identifying drug targets, bioavailability studies, *etc.*
5. In clinical studies, such as the identification of drug abuse and drug metabolites in blood, urine, saliva, *etc.*
6. In environmental monitoring and analysis, such as soil, water, and air pollutants, pesticides, water quality, food contaminants, *etc.*
7. In quality control in the chemical and petrochemical industry.

8. In geochemistry, such as in age determination, soil, rock composition, oil and gas surveying, *etc.*

8. Difference between NMR Spectroscopy and Mass Spectrometry

- NMR spectroscopy and mass spectrometry are complementary techniques. While MS can tell the mass (and thus the molecular formula) of a molecule, NMR can differentiate between structural isomers and provide information about the structural arrangement of atoms within a molecule.
- MS is a destructive method, whereas NMR is not. However, a much smaller amount of material is needed for MS techniques.

Index

^{1}H NMR spectroscopy 105
^{13}C NMR spectroscopy 104

A

Absorbance 20, 34
Absorption spectroscopy 8
Anti-stokes raman scattering 84
Atomic absorption spectroscopy 125
Auxochromes 18

B

Basic features 2
Bathochromic shift 16
Beer-Lambert's law 22
Binding kinetics 47
Biological processes 43
Biomolecules 30, 61
Brillouin scattering 81

C

Capillary electrophoresis 140
CD spectrometer 60
Chemical ionization 142
Chemical shift 101
Chromophores 16, 18
Circular dichroism 3, 55
Circular dichroism spectroscopy 55
Coherent anti-stokes Raman spectroscopy 87

D

Deshielding 100
Detector 26, 139
Dispersive spectrometer 75
Double-beam spectrophotometer 27
Drug design 117

E

Electromagnetic radiation 2
Electromagnetic spectrum 4
Electronic transitions 11
Electron ionization 141
Electron paramagnetic resonance 3
Electron spin resonance spectroscopy 121

Electrospray ionization 143, 144
Energy 7
Energy scheme 12
Enzyme kinetic assays 29
Excitation process 6
Experimental procedures 127
Extrinsic fluorophores 41

F

Fluorescence 3, 35
Fluorescence correlation spectroscopy 53
Fluorescence effects 86
Fluorescence lifetime 39
Fluorescence polarization 47
Fluorescence quenching 49
Fluorescence resonance energy transfer 50
Fluorescence spectrometer 44
Fluorescence spectroscopy 33
Fluorophores 40
Fourier transform infra-red (FTIR) spectrometer 76
Fourier transform raman spectrometer 88
Franck-condon principle 38
Frequency 4

G

Gas chromatography 140
Goniometer 115
Green fluorescent protein 42

H

High-pressure liquid chromatography 140
Hollow cathode lamp 126
Hypsochromic shift 16

I

Infra-red 3
Infra-red rays 11
Infra-red spectroscopy 69
Instrumentation 126
Internal conversion 35
Intersystem crossing 36
Intrinsic fluorophores 40
Ion source 139

K

Kasha's rule 38

L

Light 2
Light scattering 132
Light source 24, 45
Liquid chromatography instruments 29

M

Magnetic field 98
Magnetic resonance imaging 108
Mass spectrometry 135
Matrix-assisted laser desorption ionization 145
Microorganisms 30
Microwaves 10
Molar absorption coefficient 28
Molecular orbitals 12
Molecular processes 9
Molecular structures 108
Molecular vibrations 70
Monochromator 25, 45, 126

N

Nephelometry 131
NMR signals 103
NMR spectroscopy 118

NMR spectroscopy experiment 103
Nuclear magnetic resonance spectroscopy 95
Nuclear spin 97
Nucleic acids 18, 64

P

Particle 2
Phosphorescence 36
Photoluminescence 33
Physical limitations 13
Plane polarized light 56
Polarized light 56
Potassium iodide 50
Protein conformational stability 66
Protein folding 45
Protein-ligand binding 66
Proteins 17, 61
Protein sequence 151
Protein structures 89

Q

Quadrupole mass analyzer 146
Qualitative analysis 28
Quantitative analysis 29
Quantum yield 39

R

Radio waves 10
Raman scattering 82
Raman spectra 82
Raman spectrometer 87
Raman spectroscopy 79, 88
Rayleigh scattering 81, 82
Reactive nitrogen species 124
Relaxation processes 100
Resolution 14
Resonance Raman spectroscopy 88

S

Sample cell holder 26
Selectivity 14
Sensitivity 14
Shielding 100
Single-beam spectrophotometer 27
Single-molecule fluorescence spectroscopy 52
Single-molecule polarization modulation spectroscopy 53
Single-pair FRET 53
Solvents 20, 30
Spectrofluorometer 33
Spectrometer 13
Spectrometer instrument 13
Spectrometry 1
Spectrophotometer 24
Spectroscopic methods 1
Spin-spin interaction 99
Stimulated Raman scattering 87
Stokes Raman scattering 83
Surface-enhanced Raman spectroscopy 88
Surface-enhanced resonance Raman spectroscopy 88

T

Tandem mass spectrometry 148
Thompson scattering 82
Time-of-flight mass analyzer 147
Transmittance 20
Tryptophan fluorescence 41
Tryptophan fluorescence intensity curves 46
Turbidimetry 131

U

Ultraviolet 15
UV/visible rays 11

V

Visible spectroscopy 15

W

Wave 2
Wavenumber 4

X

X-ray crystallography 113
X-rays 11
X-ray spectroscopy 111

9 789359 191492

Printed at Repro India Ltd.